PROGRESS IN CLINICAL AND BIOLOGICAL RESEARCH

RECENT TITLES

Please contact publisher for information about previous titles in this series.

HEMOGLOBIN SWITCHING
Part A: Transcriptional Regulation

HEMOGLOBIN SWITCHING

Part A: Transcriptional Regulation

Proceedings of the Sixth Conference on Hemoglobin Switching held in Airlie, Virginia, September 24–27, 1988

Editors

George Stamatoyannopoulos
Division of Medical Genetics
Department of Medicine
University of Washington
Seattle, Washington

Arthur W. Nienhuis
Clinical Hematology Branch
National Heart, Lung, and Blood Institute
National Institutes of Health
Bethesda, Maryland

ALAN R. LISS, INC. • NEW YORK

Library of Congress Cataloging-in-Publication Data

Conference on Hemoglobin Switching (6th : 1988 :
    Airlie, Va.)
    Hemoglobin switching.

    (Progress in clinical and biological research ;
v. 316)
    Includes bibliographies and index.
    Contents: pt. A. Transcriptional regulation --
pt. B. Cellular and molecular mechanisms.
    1. Hemoglobin--Synthesis--Congresses. 2. Hemo-
globin polymorphisms--Congresses. 3. Globin genes--
Congresses. I. Stamatoyannopoulos, George.
II. Nienhuis, Arthur W. III. Title. IV. Series.
[DNLM: 1. Gene Expression Regulation--congresses.
2. Globin--genetics--congresses. 3. Transcription
Factors--congresses. W1 PR668E v.316 /
WH 190 C7475h 1988]
QP96.5.C67 1989   599'.0113   89-35204
ISBN 0-8451-5166-5 (set)
ISBN 0-8451-5402-8 (pt. A)
ISBN 0-8451-5403-6 (pt. B)

Contents

TRANSCRIPTIONAL CONTROL OF β- AND γ-GLOBIN GENES

TRANSCRIPTIONAL CONTROL: HPFH MUTANTS

OTHER ERYTHROID

Contributors

Santina Acuto, Departments of Genetics and Development, and Medicine, Columbia University, New York, NY 10032; present address: Department of Hematology, Osp. v. Cervello, CFRS, Palermo, Sicily, Italy [163]

Mike Antoniou, Laboratory of Gene Structure and Expression, National Institute for Medical Research, London NW7 1AA, England [1,37]

Arthur Bank, Departments of Genetics and Development, and Medicine, Columbia University, New York, NY 10032 [163]

Kerry M. Barnhart, DeWitt Wallace Research Laboratory and the Graduate Program in Molecular Biology, Memorial Sloan-Kettering Cancer Center, New York, NY 10021 [343]

David P. Bazett-Jones, Department of Medical Biochemistry, University of Calgary, Calgary T2N 4N1, Alberta, Canada [291]

Denise Beaupain, INSERM U 91, Hôpital Henri Mondor, 94010 Créteil, France [359]

Richard R. Behringer, Laboratory of Reproductive Physiology, School of Veterinary Medicine, University of Pennsylvania, Philadelphia, PA 19104 [47]

Patricia E. Berg, Laboratory of Chemical Biology, National Institute of Diabetes, and Digestive and Kidney Diseases, National Institutes of Health, Bethesda, MD 20892 [193]

Kristina L. Blanchard, Howard Hughes Medical Institute, University of Michigan, Ann Arbor, MI 48109 [129]

David M. Bodine, Clinical Hematology Branch, National Heart, Lung, and Blood Institute, National Institutes of Health, Bethesda, MD 20892 [149]

Alison Brewer, Department of Biophysics, Cell and Molecular Biology, University of London, King's College, London WC2B 5RL, England [105]

Ralph L. Brinster, Laboratory of Reproductive Physiology, School of Veterinary Medicine, University of Pennsylvania, Philadelphia, PA 19104 [47]

Mark L. Brown, Department of Medical Biochemistry, University of Calgary, Calgary T2N 4N1, Alberta, Canada [291]

Shi Xian Cao, Laboratory of Chemical Biology, National Institute of Diabetes, and Digestive and Kidney Diseases, National Institutes of Health, Bethesda, MD 20892 [279]

The numbers in brackets are the opening page numbers of the contributors' articles.

Francoise Catala, Laboratory of Gene Structure and Expression, National Institute for Medical Research, London NW7 1AA, England **[1]**

Resy Cavallesco, Harvard-MIT Division of Health Sciences and Technology, and Department of Biology, Massachusetts Institute of Technology, Cambridge, MA 02139 **[63]**

Claude Chabret, INSERM U 91, Hôpital Henri Mondor, 94010 Créteil, France **[359]**

G.K.T. Chan, Department of Medical Biochemistry, University of Calgary, Calgary T2N 4N1, Alberta, Canada **[291]**

Ok-Ryun B. Choi, Department of Biochemistry, Molecular Biology and Cell Biology, Northwestern University, Evanston, IL 60208 **[89]**

Roger B. Cohen, National Heart, Lung, and Blood Institute, National Institutes of Health, Bethesda, MD 20892 **[193]**

Francis S. Collins, Departments of Internal Medicine, Human Genetics, and Howard Hughes Medical Institute, University of Michigan, Ann Arbor, MI 48109 **[129]**

Phil Collis, Laboratory of Gene Structure and Expression, National Institute for Medical Research, London NW7 1AA, England **[37]**

Frank Costantini, Department of Genetics and Development, College of Physicians and Surgeons, Columbia University, New York, NY 10032 **[203]**

Alison Cowie, Departments of Physiology, Biochemistry, and Biophysics, University of California at San Francisco, San Francisco, CA 94143 **[117]**

Harish P.G. Dave, Laboratory of Chemical Biology, National Institute of Diabetes, and Digestive and Kidney Diseases, National Institutes of Health, Bethesda, MD 20892 **[279]**

Ann Dean, Laboratory of Cellular and Developmental Biology, National Institute of Diabetes, and Digestive and Kidney Diseases, National Institutes of Health, Bethesda, MD 20892 **[179]**

Ernie deBoer, Laboratory of Gene Structure and Expression, National Institute for Medical Research, London NW7 1AA, England **[1, 359]**

Joseph DeSimone, Departments of Medicine and Genetics, University of Illinois College of Medicine, and the VA Westside Medical Center, Chicago, IL 60612 **[261]**

Niall Dillon, Laboratory of Gene Structure and Expression, National Institute for Medical Research, London NW7 1AA, England **[37]**

Maryann Donovan-Peluso, Department of Genetics and Development, Columbia University, New York, NY 10032 **[163]**

Anne Dubart, INSERM U 91, Hôpital Henri Mondor, 94010 Créteil, France **[359]**

Debra J. Endean, Department of Biochemistry, Molecular Biology and Cell Biology, Northwestern University, Evanston, IL 60208 **[89]**

James Douglas Engel, Department of Biochemistry, Molecular Biology and Cell Biology, Northwestern University, Evanston, IL 60208 **[89]**

Tariq Enver, Department of Biophysics, Cell and Molecular Biology, University of London, King's College, London WC2B 5RL, England; present address: Division of Medical Genetics, Department of Medicine, University of Washington, Seattle, WA 98195 **[105]**

Elliot Epner, Fred Hutchinson Cancer Research Center and the University of Washington School of Medicine, Seattle, WA 98104 **[15]**

T. Evans, Laboratory of Molecular Biology, National Institute of Diabetes, and Digestive and Kidney Diseases, National Institutes of Health, Bethesda, MD 20892 **[73]**

G. Felsenfeld, Laboratory of Molecular Biology, National Institute of Diabetes, and Digestive and Kidney Diseases, National Institutes of Health, Bethesda, MD 20892 **[73]**

Kevin P. Foley, Department of Biochemistry, Molecular Biology and Cell Biology, Northwestern University, Evanston, IL 60208 **[89]**

Bernard G. Forget, Departments of Internal Medicine and Human Genetics, Yale University School of Medicine, New Haven, CT 06510 **[247]**

William C. Forrester, Fred Hutchinson Cancer Research Center and the University of Washington School of Medicine, Seattle, WA 98104 **[15]**

Karin Gaensler, Departments of Physiology, Biochemistry, and Biophysics, University of California at San Francisco, San Francisco, CA 94143 **[117]**

James L. Gallarda, Department of Biochemistry, Molecular Biology and Cell Biology, Northwestern University, Evanston, IL 60208 **[89]**

Barbara Giglioni, Centro per lo Studio della Patologia Cellulare del Consiglio Nazionale Ricerche, Milano 20133, Italy **[229]**

Frances P. Gillespie, Department of Internal Medicine, Yale University School of Medicine, New Haven, CT 06510 **[247]**

John Gilman, Department of Cell and Molecular Biology, Medical College of Georgia, Augusta, GA 30812-2100 **[229]**

Qi-Hui Gong, Laboratory of Cellular and Developmental Biology, National Institute of Diabetes, and Digestive and Kidney Diseases, National Institutes of Health, Bethesda, MD 20892 **[179]**

Todd A. Gray, Department of Human Genetics, University of Michigan, Ann Arbor, MI 48109 **[129]**

David R. Greaves, Laboratory of Gene Structure and Expression, National Institute for Medical Research, London NW7 1AA, England **[37]**

Frank Grosveld, Laboratory of Gene Structure and Expression, National Institute for Medical Research, London NW7 1AA, England **[1, 37, 359]**

Mark Groudine, Fred Hutchinson Cancer Research Center and the University of Washington School of Medicine, Seattle, WA 98104 **[15]**

Deborah L. Gumucio, Department of Internal Medicine, University of Michigan, Ann Arbor, MI 48109 **[129]**

Pablo D. Gutman, Laboratory of Chemical Biology, National Institute of Diabetes, and Digestive and Kidney Diseases, National Institutes of Health, Bethesda, MD 20892 **[279]**

Olivia Hanscombe, Laboratory of Gene Structure and Expression, National Institute for Medical Research, London NW7 1AA, England **[37]**

Grant Hartzog, Departments of Physiology, Biochemistry, and Biophysics, University of California at San Francisco, San Francisco, CA 94143 **[117]**

P. Hogben, Sir William Dunn School of Pathology, University of Oxford, Oxford OX1 3RE, England **[323]**

Anna Hom, Department of Genetics and Development, Columbia University, New York, NY 10032 **[163]**

George Huang, Harvard-MIT Division of Health Sciences and Technology, and Department of Biology, Massachusetts Institute of Technology, Cambridge, MA 02139 **[63]**

Jacky Hurst, Laboratory of Gene Structure and Expression, National Institute for Medical Research, London NW7 1AA, England **[37]**

P.D. Jackson, Laboratory of Molecular Biology, National Institute of Diabetes, and Digestive and Kidney Diseases, National Institutes of Health, Bethesda, MD 20892 **[73]**

Russel E. Kaufman, Departments of Medicine/Hematology and Biochemistry, Duke University Medical Center, Durham, NC 27710 **[237]**

N. Kawamura, Department of Medical Biochemistry, University of Calgary, Calgary T2N 4N1, Alberta, Canada **[291]**

James Kaysen, Department of Genetics and Development, Columbia University, New York, NY 10032 **[163]**

Chul G. Kim, DeWitt Wallace Research Laboratory and the Graduate Program in Molecular Biology, Memorial Sloan-Kettering Cancer Center, New York, NY 10021 **[343]**

J. Knezetic, Laboratory of Molecular Biology, National Institute of Diabetes, and Digestive and Kidney Diseases, National Institutes of Health, Bethesda, MD 20892 **[73]**

Peter Lamb, Sir William Dunn School of Pathology, University of Oxford, Oxford OX1 3RE, England **[269, 323]**

Donald E. Lavelle, Departments of Medicine and Genetics, University of Illinois College of Medicine, and the VA Westside Medical Center, Chicago, IL 60612 **[261]**

Richard F. Lee, Department of Molecular Genetics, Biochemistry and Microbiology, University of Cincinnati, Cincinnati, OH 45267-0524; present address: The Molecular Diagnostics Center, Good Samaritan Hospital, Cincinnati, OH 45220 **[139]**

Mark Leonard, Department of Biophysics, Cell and Molecular Biology, University of London, King's College, London WC2B 5RL, England **[105]**

C. Lewis, Laboratory of Molecular Biology, National Institute of Diabetes, and Digestive and Kidney Diseases, National Institutes of Health, Bethesda, MD 20892 **[73]**

Timothy Ley, Department of Medicine, Washington University School of Medicine, St. Louis, MO; present address: Division of Hematology/Oncology, Jewish Hospital, St. Louis, MO 63110 **[149]**

Henry Lin, Clincal Hematology Branch, National Heart, Lung, and Blood Institute, National Institutes of Health, Bethesda, MD 20892 **[149]**

Michael Lindenbaum, Laboratory of Gene Structure and Expression, National Institute for Medical Research, London NW7 1AA, England **[37]**

Jerry B. Lingrel, Department of Molecular Genetics, Biochemistry and Microbiology, University of Cincinnati, Cincinnati, OH 45267-0524 **[139]**

Joyce A. Lloyd, Department of Molecular Genetics, Biochemistry and Microbiology, University of Cincinnati, Cincinnati, OH 45267-0524 **[139]**

L. Locklear, Department of Medical Biochemistry, University of Calgary, Calgary T2N 4N1, Alberta, Canada **[291]**

Irving M. London, Harvard-MIT Division of Health Sciences and Technology, and Department of Biology, Massachusetts Institute of Technology, Cambridge, MA 02139 **[63]**

Christopher Lowrey, Clinical Hematology Branch, National Heart, Lung, and Blood Institute, National Institutes of Health, Bethesda, MD 20892 **[149]**

Jeanne Magram, Department of Genetics and Development, College of Physicians and Surgeons, Columbia University, New York, NY 10032; present address: Department of Microbiology and Immunology, University of California at San Francisco Medical Center, San Francisco, CA 94143 **[203]**

N. Malgaretti, Dipartimento di Genetica e di Biologia dei Microrganismi, Università di Milano, Milano 20133, Italy **[229]**

Avgi Mamalaki, Institute of Molecular Biology and Biotechnology, Foundation for Research and Technology, Hellas, Heraklion 711 10, Crete, Greece **[335]**

R. Mantovani, Dipartimento di Genetica e di Biologia dei Microrganismi, Università di Milano, Milano 20133, Italy **[229]**

David I.K. Martin, Department of Hematology/Oncology, Children's Hospital and Howard Hughes Medical Institute, Boston, MA 02115 **[217]**

Kevin T. McDonagh, Clinical Hematology Branch, National Heart, Lung, and Blood Institute, National Institutes of Health, Bethesda, MD 20892 **[149]**

Anil G. Menon, Department of Molecular Genetics, Biochemistry and Microbiology, University of Cincinnati, Cincinnati, OH 45267-0524; present address: W. Laboratory of Neurogenetics, Massachusetts General Hospital, Boston, MA 02114 **[139]**

James E. Metherall, Department of Human Genetics, Yale University School of Medicine, New Haven, CT 06510; present address: Department of Molecular Genetics, University of Texas Southwestern Medical Center at Dallas, Dallas, TX 75235 **[247]**

Vincent Mignotte, INSERM U 91, Hôpital Henri Mondor, 94010 Créteil, France **[359]**

Moshe Mittelman, Laboratory of Chemical Biology, National Institute of Diabetes, and Digestive and Kidney Diseases, National Institutes of Health, Bethesda, MD 20892 **[193]**

Nicholas K. Moschonas, Institute of Molecular Biology and Biotechnology, Foundation for Research and Technology, Hellas, Department of Biology, University of Crete, Heraklion 711 10, Crete, Greece **[335]**

Richard M. Myers, Departments of Physiology, Biochemistry, and Biophysics, University of California at San Francisco, San Francisco, CA 94143 **[117]**

J. Nickol, Laboratory of Molecular Biology, National Institute of Diabetes, and Digestive and Kidney Diseases, National Institutes of Health, Bethesda, MD 20892 **[73]**

S. Nicolis, Dipartimento di Genetica e di Biologia dei Microrganismi, Università di Milano, Milano 20133, Italy **[229]**

Karen Niederreither, Department of Genetics and Development, College of Physicians and Surgeons, Columbia University, New York, NY 10032; present address: Graduate Program in Cancer Biology, M.D. Anderson Hospital, Houston, TX 77030 **[203]**

Arthur W. Nienhuis, Clinical Hematology Branch, National Heart, Lung, and Blood Institute, National Institutes of Health, Bethesda, MD 20892 **[149]**

Constance Tom Noguchi, Laboratory of Chemical Biology, National Institute of Diabetes, and Digestive and Kidney Diseases, National Institutes of Health, Bethesda, MD 20892 **[301, 313]**

Ulrike Novak, Fred Hutchinson Cancer Research Center and the University of Washington School of Medicine, Seattle, WA 98104 **[15]**

David O'Neill, Department of Pathology, Columbia University, New York, NY 10032 **[163]**

Stuart H. Orkin, Department of Hematology/Oncology, Children's Hospital and Howard Hughes Medical Institute, Boston, MA 02115 **[217]**

Sergio Ottolenghi, Dipartimento di Genetica e di Biologia dei Microrganismi, Università di Milano, Milano 20133, Italy **[229]**

Richard D. Palmiter, Department of Biochemistry, Howard Hughes Medical Institute, University of Washington, Seattle, WA 98195 **[47]**

Roger Patient, Department of Biophysics, Cell and Molecular Biology, University of London, King's College, London WC2B 5RL, England **[105]**

Carlos Perez-Stable, Department of Genetics and Development, College of Physicians and Surgeons, Columbia University, New York, NY 10032 **[203]**

Nicholas J. Proudfoot, Sir William Dunn School of Pathology, University of Oxford, Oxford OX1 3RE, England **[269, 323]**

Mary Purucker, Clinical Hematology Branch, National Heart, Lung, and Blood Institute, National Institutes of Health, Bethesda, MD 20892 **[149]**

Ruo-Lan Qian, Laboratory of Chemical Biology, National Institute of Diabetes, and Digestive and Kidney Diseases, National Institutes of Health, Bethesda, MD 20892 **[193]**

Natacha Raich, INSERM U 91, Hôpital Henri Mondor, 94010 Créteil, France **[359]**

M. Reitman, Laboratory of Molecular Biology, National Institute of Diabetes, and Digestive and Kidney Diseases, National Institutes of Health, Bethesda, MD 20892 **[73]**

Marc Romana, INSERM U 91, Hôpital Henri Mondor, 94010 Créteil, France **[359]**

Paul-Henri Romeo, INSERM U 91, Hôpital Henri Mondor, 94010 Créteil, France **[359]**

A. Ronchi, Dipartimento di Genetica e di Biologia dei Microrganismi, Università di Milano, Milano 20133, Italy **[229]**

Kirsten L. Rood, Department of Human Genetics, University of Michigan, Ann Arbor, MI 48109 **[129]**

Thomas M. Ryan, Department of Biochemistry, Schools of Medicine and Dentistry, University of Alabama at Birmingham, Birmingham, AL 35294 **[47]**

Alan N. Schechter, Laboratory of Chemical Biology, National Institute of Diabetes, and Digestive and Kidney Diseases, National Institutes of Health, Bethesda, MD 20892 **[193, 279]**

Michael Sheffery, DeWitt Wallace Research Laboratory and the Graduate Program in Molecular Biology, Memorial Sloan-Kettering Cancer Center, New York, NY 10021 **[343]**

William B. Solomon, Harvard-MIT Division of Health Sciences and Technology, and Department of Biology, Massachusetts Institute of Technology, Cambridge, MA 02139; present address: Division of Oncology and Morse Institute of Molecular Biology, State University of New York, Brooklyn, NY 11203 **[63]**

Laura Stuve, Departments of Physiology, Biochemistry, and Biophysics, University of California at San Francisco, San Francisco, CA 94143 **[117]**

Kathryn Sykes, Department of Biochemistry, Duke University Medical Center, Durham, NC 27710 **[237]**

Dale Talbot, Laboratory of Gene Structure and Expression, National Institute for Medical Research, London NW7 1AA, England **[37]**

Tim M. Townes, Department of Biochemistry, Schools of Medicine and Dentistry, University of Alabama at Birmingham, Birmingham, AL 35294 **[47]**

Dorothy Y.H. Tuan, Harvard-MIT Division of Health Sciences and Technology, and Department of Biology, Massachusetts Institute of Technology, Cambridge, MA 02139 **[63]**

Greet Blom van Assendelft, Laboratory of Gene Structure and Expression, National Institute for Medical Research, London NW7 1AA, England **[37]**

Yuko Wada, Laboratory of Chemical Biology, National Institute of Diabetes, and Digestive and Kidney Diseases, National Institutes of Health, Bethesda, MD 20892 **[301]**

Lee Wall, Laboratory of Gene Structure and Expression, National Institute for Medical Research, London NW7 1AA, England **[1,359]**

Maggie Walmsley, Department of Biophysics, Cell and Molecular Biology, University of London, King's College, London WC2B 5RL, England **[105]**

Paul Watt, Sir William Dunn School of Pathology, University of Oxford, Oxford OX1 3RE, England **[269]**

E. Whitelaw, Sir William Dunn School of Pathology, University of Oxford, Oxford OX1 3RE, England **[323]**

Donna M. Williams, Laboratory of Chemical Biology, National Institute of Diabetes, and Digestive and Kidney Diseases, National Institutes of Health, Bethesda, MD 20892 **[193]**

Angus Wilson, Department of Biophysics, Cell and Molecular Biology, University of London, King's College, London WC2B 5RL, England **[105]**

Yongji Wu, Laboratory of Chemical Biology, National Institute of Diabetes, and Digestive and Kidney Diseases, National Institutes of Health, Bethesda, MD 20892 **[313]**

Zhuoying Yang, Department of Biochemistry, Molecular Biology and Cell Biology, Northwestern University, Evanston, IL 60208 **[89]**

Contents of Part B: Cellular and Molecular Mechanisms

INSIGHTS FROM HEMOPOIETIC CELL LINES

GLOBIN GENE TRANSFER

MODULATION OF HbF AND GLOBIN GENE SWITCHING

PREFACE

In organizing the First Conference on Hemoglobin Switching in 1978, we invited every investigator working directly on hemoglobin switching and those engaged in the effort to characterize globin genes using the new tools of molecular biology. This meeting took place in the Batelle Institute in Seattle and included only 75 participants. One third of the papers were devoted to description of developmental switches in vivo. The use of clonogenic cultures for analysis of the mechanism of regulation of erythropoiesis and of fetal hemoglobin synthesis also received considerable emphasis. Most of the molecular work presented pertained to RNA structure and metabolism and chromatin structure. Presented also was the first information about the genomic map of the human gamma globin genes, as obtained by Southern blot analysis, and reports of the recently cloned rabbit, mouse and human β globin genes.

The last 10 years has witnessed extraordinary progress in the field of hemoglobin switching. A major emphasis at the Sixth Conference on Hemoglobin Switching was on the detailed mechanisms involved in the transcriptional control of globin genes. Descriptive studies of hemoglobin switching have given way to pharmacological experiments designed to investigate the mechanism of induction of fetal hemoglobin in vivo. Therapeutic efforts to enhance fetal hemoglobin synthesis in patients with sickle cell disease are now a reality.

The Sixth Conference on Hemoglobin Switching was held from September 24-27, 1988 in Airlie House, Virginia. Approximately 200 investigators attended; the size of the Conference was limited by available facilities. Most speakers and discussants had not attended the First Conference on Hemoglobin Switching but rather were newcomers to this field during the past 10 years.

The Proceedings of the Sixth Conference are presented in two volumes. The first volume is a comprehensive report of current work pertaining to the transcriptional control of globin gene expression. One of the remarkable recent discoveries is the existence of the locus activating or dominant control region upstream from the ϵ globin gene and its role in establishing position independent, copy number dependent gene expression in cultured cells and transgenic animals. Efforts to identify the DNA sequences involved in the function of this region are well underway as reflected in the five papers devoted to this topic. Structural and functional characterization of several other cis acting control elements is also well advanced; remarkable are the number and complexity of protein interactions that have been defined. One erythroid specific protein has been identified and there may be others. Much remains to be accomplished before the exact molecular mechanisms of hemoglobin switching are defined but the experiments to be performed are apparent. One may be encouraged by the enthusiastic effort of the many young investigators who have been attracted to this area of investigation.

Exciting among the results reported in the second volume are the effect of various treatments designed to enhance fetal hemoglobin synthesis. Although such therapy remains experimental, hematological improvement in sickle cell patients treated with hydroxyurea is becoming evident. Several other drugs and hematopoietic growth factors have been shown to stimulate HbF synthesis alone and in various combinations. The opportunity for careful clinical investigation, if aggressively pursued, should result in real patient benefit within the next several years.

Presented at the Conference and included in the Proceedings are summaries of recent progress in the analysis of the structure and organization of globin genes. Pertinent features of chromatin structure and interaction of DNA sequences with the nuclear matrix are also described. Several new mutants are reported that provide additional insights into globin gene regulation.

The Conference was sponsored by the National Institute of Diabetes, Digestive Disease and Kidney

Diseases, the National Heart, Lung and Blood Institute, and the National Cancer Institute. The following corporations provided funds that allowed support of invited young investigators and post-doctoral fellows (number of persons supported is shown in parentheses): Genetics Institute (one), Amgen, Inc. (one), Sandoz Research Institute (four), Sandoz Pharmaceuticals (two), Burroughs Wellcome Company (one), Ross Laboratories (one).

The Editors wish to thank Drs. David Badman and Clarice Reid for their encouragement and support.

G. Stamatoyannopoulos
A.W. Nienhuis

Hemoglobin Switching, Part A: Transcriptional Regulation, pages 1–13
© 1989 Alan R. Liss, Inc.

THE REGULATION OF THE HUMAN γ- AND β-GLOBIN DOMAIN

Lee Wall, Francoise Catala, Mike Antoniou,
Ernie deBoer and Frank Grosveld
Laboratory of Gene Structure and Expression,
National Institute for Medical Research,
The Ridgeway, Mill Hill, London NW7 1AA UK.

INTRODUCTION

The human γ and β-globin genes are part of a multigene family that is expressed in a developmental and stage specific manner (for review, see Collins & Weissman, 1984). The γ-globin genes are normally expressed during the foetal stage of development, when the liver is the hematopoeitic organ. Around the time of birth, the expression of these genes diminishes and is replaced by expression of δ- and β-globin genes in the adult bone marrow. The entire cluster of genes is regulated by a region at the border of the globin domain (Grosveld et al., 1987), which is characterized by a set of DNaseI hypersensitive sites (Tuan et al., 1985; Forrester et al., 1987; Grosveld et al., 1987). This dominant control region (DCR) is completely erythroid specific but does not appear to have any particular stage specificity for the different globin genes (Dillon et al., unpublished). Instead, the expression of the individual genes appears to be regulated by sequences immediately flanking the genes. There is, however, some co-ordination competition between the genes within the entire domain, as demonstrated by the elevated levels of γ-globin expression in several different disorders (for review, see Poncz et al., 1988).

REGULATORY REGIONS FLANKING THE HUMAN β-GLOBIN GENES

The β-globin gene and its immediate flanking regions contain a set of DNaseI hypersensitive sites in vivo

(Groudine *et al.*, 1983) which characterize the position of two enhancer elements and a promoter region. These are regulated by the action of trans-acting protein factors (Baron and Maniatis, 1986; Wrighton and Grosveld, 1988), possibly without the need for a round of DNA replication. The enhancers have been characterized in detail, both functionally and biochemically (Fig. 1). The first enhancer is located at the border of the second intervening sequence and the third exon (Behringer *et al.*, 1987; Antoniou *et al.*, 1988; Wall *et al.*, unpublished) and contains at least three binding sites for an erythroid specific protein (see below). The second enhancer is located 600bp downstream of the gene (Behringer *et al.*, 1987; Kollias *et al.*, 1987; Trudel *et al.*, 1987; Antoniou *et al.*, 1988) and its protein binding sites have been characterized in detail (Wall *et al.*, 1988). Both enhancer elements are active in transgenic mice and MEL cells. The addition of the β-globin gene enhancer to the γ-globin gene switches its expression from strictly embryonic in the mouse to the embryonic and foetal/adult stages (Behringer *et al.*, 1987; Kollias *et al.*, 1986, 1987; Trudel *et al.*, 1987). The β-globin enhancers are not only developmentally specific, but they are also only active in erythroid cells (Antoniou *et al.*, 1988 and unpublished). Like classical enhancers, they act in an orientation, position and (relatively) distance-independent manner on homologous and at least some heterologous genes. There has been one report of an enhancer sequence flanking the 3' side of the $^{A}\gamma$ globin gene (Bodine *et al.*, 1987) which would not appear to be tissue-specific. Its presence, however, has yet to be confirmed by other laboratories. Interestingly, there are no in vivo data to indicate the importance of either of the γ- or β- 3' flanking enhancers. In other words, mutations in all the different elements affecting all steps in the transcription (and translation) of the β-globin gene have been found (promoter, splicing, polyA, etc.), with the exception of the 3' flanking enhancer. It is, therefore, possible that such mutations do not exist, or that they exist but are unimportant (and therefore not picked up by patient screening).

The promoters of the γ- and β-globin genes are also regulated (Wright *et al.*, 1984; Charnay *et al.*, 1984; Kioussis *et al.*, 1985; Rutherford and Nienhuis, 1987) and both of these have now been characterized in some detail.

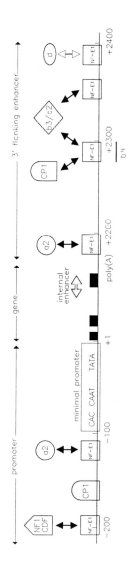

Fig. 1:   Schematic representation of the factors binding to the human -globin promoter (deBoer et al., 1988) and 3' enhancer (Wall et al., 1988).

The promoter and enhancer are on the same scale, the gene and immediate 3' flanking region have been compressed to a smaller scale. The symbols indicate the various protein factors, CDF, NF1, GF1, CP1, a2 and b3/c2 (Wall, et al., 1988). The factors binding to the minimal promoter (TATA, CAAT and CAC boxes) have not been indicated. Solid double arrows indicate binding of different factors to the same or overlapping sites.

The promoter region, particularly of the β-globin gene, has been analyzed in great detail. The (minimal) promoter region, up to 100bp 5' to the transcription initiation start (-100) has been shown to contain the (-30) TATA box, (-70) CAAT box and (-90) CAC box motifs which are necessary for efficient transcription (Myers et al., 1986; Dierks et al., 1983; Grosveld et al.,1982). Interestingly, the minimal promoter of the human β-globin gene does not appear to be regulated upon the induction of MEL cells (Antoniou et al., 1988), in contrast to the promoter of the mouse β-globin gene (Cowie and Myers, 1988). It has even been suggested that only part of the minimal promoter might be important, since a deleted (-48) promoter was still expressed specifically in erythroid cells in the presence of the enhancers (Townes et al., 1985). However, transfection experiments indicate that the promoter does contain specific regulatory elements which are required for the induction in MEL cells (Wright et al., 1984; Anagnou et al., 1986) and that these are located outside the minimal promoter (Antoniou et al., 1988). The in vivo evidence for the importance of the different promoter sequences is fairly limited; only two mutations in the minimal promoter have been found (TATA box and CAC box) (Antonarakis et al., 1984; Orkin et al., 1982, 1983; Treisman et al., 1983; Orkin et al., 1984). Deletion analysis of the γ-globin gene promoter also shows that the region outside the minimal promoter is important for its regulated expression (Rutherford and Nienhuis, 1987). Our recent data show that sequential deletion of 5' sequences allows distinction of at least three functional regions outside the minimal promoter sequence (Catala et al., 1988). These data agree with the in vivo mutations found in non-deletion HPFH, most of which fall in the same regions (for review, see Poncz et al., 1988) and which affect the binding of particular trans-acting factors (Mantovani et al., 1988 and below).

We have recently analyzed the protein factors that bind to the β-globin promoter and 3' flanking enhancer, using nuclear protein extracts and in vitro DNA binding experiments. The results of this are summarized in Fig. 1. The 3' enhancer (250bp PstI fragment) and the promoter (up to -250) contain four DNA binding sites each, for a protein we have called NF-E1. This protein is very similar between mouse, man and chicken (Wall et al., 1988; Evans et al., 1988) and binds to all erythroid promoters

and globin enhancers that have been analyzed to date. It
is a 37-40Kd erythroid specific protein present at all
stages of development and has been purified in several
laboratories. Although it is involved in a positive
stimulation of transcription (Ottolenghi, 1988; Mignotte
et al., unpublished and see below), the presence of this
factor alone, unless it is altered in some way, is not
sufficient for this effect, because it is present before
and after the induction of MEL cells. Each of the NF-El
binding sites in the enhancer is also capable of binding a
number of other factors. The first binds a ubiquitous
factor a2 and is also found in the promoter region   (at
-120). The binding efficiencies for a2 and NF-El are
reversed at the two sites (-120 binds a2 better) and the
ratio of a2 to NF-El is significantly higher in K562 (non-
β expressing) than in MEL cells (Wall et al., 1988; deBoer
et al., 1988). Although a2 and NF-El have a different
binding site, they bind competitively to the same region
and it is therefore conceivable that a2 may play an active
role in the suppression of β-globin transcription (Wall
and Antoniou, unpublished). The second binding site in
the enhancer is very complicated. It has a low affinity
for NF-El, binds two ubiquitous factors, the CAAT box
binding factor CP1 (Chodosh et al.,1988) and a factor we
termed b3/c2 (Wall et al., 1988), which appears to be
related to an octamer binding protein (Wall et al.,
unpublished). Lastly, a second erythroid specific factor,
b4, binds to this region. From competition experiments,
it initially appeared to consist of a complex of NF-El
with a second protein, however, fractionation indicates it
to be a separate factor (Wall and deBoer, unpublished).
The third site also binds b3/c2 competitively at a site
which overlaps the 3' side of the NF-El binding site (Wall
et al., 1988). At the fourth site a different erythroid
specific protein, d2, can also bind, it is more abundant
in K562 non β-globin expressing erythroid cells and might
(like a2) be involved in the suppression of β-globin
expression in these cells. The internal enhancer has been
analyzed in much less detail, but it is clear that it has
at least three binding sites for the factor NF-El (Wall et
al., unpublished). These three sites span the EcoRI site
in the third exon of the β-globin gene in agreement with
the mapping data for this enhancer (Behringer et al.,
1987). In the promoter region NF-El binds at three
positions, at the CAAT box and the -120 region with low
affinity and at the -200 region with high affinity. The

CAAT box also binds CP1 and the -120 region binds the protein a2 (deBoer *et al.*, 1988 and see above). Immediately upstream of the -200 NF-E1 sites, another non-erythroid specific complex is found, which appears to be related to NF1 (Jones *et al.*, 1987) and CDP (Superti-Furga *et al.*, 1988) as defined by competition experiments (deBoer *et al.*, 1988). The CAAT box and -120 region span the CAC box, which is known to be required for the efficient functioning of the promoter in vivo (Orkin *et al.*, 1982; 1984; Treisman *et al.*, 1983; Myers *et al.*, 1986). It binds Sp1 with low affinity and a second CAC box specific factor similar to OGT2 (Xiao *et al.*, 1987; Davidson *et al.*, 1988) with very high affinity (Spanopoulou, unpublished). Presumably the effect of the upstream promoter is mediated to the minimal promoter (CAAT box, TATA box) by the binding of this protein in a similar fashion as described by Schule *et al.* (1988) who described co-operativity between the CAC box binding factor and the glucacorticoid receptor. The -120 and -200 NF-E1 sites span a region at -150 which binds two non-erythroid complexes CP1 and with low affinity NF1 (deBoer *et al.*, 1988). This region had previously been identified by deletion experiments to be important for the induction of the β-globin promoter in differentiating MEL cells (Antoniou *et al.*, 1988). We therefore tested which of the upstream regions (-120, -150, -200) are required for this effect by linking different combinations of these elements to the minimal promoter (Fig. 2). Analysis of the RNA levels shows that the minimal promoter is only inducible in MEL cells when combined with the -200 and -150 regions or the -150 and -120 regions (lanes 103-120, 140-400+ (-150) and -184), but not with the -120, -150 or -200 regions alone (lanes -103-164 and -138, or 103-400+ (100) or -140-400). This confirms that the -150 region is crucial for the induction of the promoter but that protein-protein interactions in the upstream part of the promoter (probably between CP1 and NF-E1) are important in this process. It does, however, not explain how this is mediated, since there is no detectable change in any of the factors before and after induction of the cells. Moreover, all these factors are also present in K562 cells which do not express the β-globin gene and apart from the a2/NF-E1 quantitative difference, i.e. the analysis of DNA binding activities has not so far provided a simple explanation for the stage specificity of the promoter (Antoniou *et al.*, 1988).

THE PROMOTER OF THE HUMAN γ-GLOBIN GENE

The γ-globin promoter has a different composition of protein binding sites from the β-globin promoter, but uses mostly the same proteins (Fig. 3). A series of progressive deletions of the promoter of the γ-globin indicate several regions of importance (Catala et al., 1988). From a deletion at -400 to a deletion at -260 shows an increase of expression in K562 cells when the $^A\gamma$-gene is part of a globin minilocus (see below). Further deletions to -160 lead to a decrease to 0.5 of the activity obtained with the -260 promoter. Deletion (to -122) of the CAC box leads to another 5-fold loss of activity, but leaves a clearly detectable level of transciption for the promoter containing only the duplicated CAAT box region and the TATA box (Catala et al., 1988).

The duplicated CAAT box binds three factors: CP1 which interacts preferentially with the proximal CAAT box, NF-E1 which binds with higher affinity to the distal CAAT region and the CAAT displacement protein CDP which recognizes sequences in both CAAT box regions (Superti-Furga et al., 1988). Interestingly, these authors also showed that the Greek HPFH mutation in the distal CAAT box region (G->A at -117) increases the binding of CP1 and CDP, but strongly reduces the binding of NF-E1 in vitro. From this result they postulate that NF-E1 may play a role in the suppression of the γ-globin gene in adult erythroid cells. In contrast to this, it has also been reported that a second erythroid specific factor (present in K562 cells) binds to the distal CAAT region and that this factor is involved in the HPFH phenotype (Ottolenghi, 1988). Further upstream the CAC box binds a similar factor as the β-globin gene (Mantovani et al., 1988; Catala et al., 1988) which is clearly required for the efficient function of the promoter (Catala et al., 1988). At the -175 region two factors are bound, NF-E1 at -184 which binds competitively but with higher affinity than ubiquitous octamer binding factor (OTF1) at -175 (Fletcher et al., 1987; Mantovani et al., 1988; Catala et al., 1988). Functional experiments using point mutations (including the non-deletion HPFH -157 T to C change) indicates that the octamer factor may play no role at all

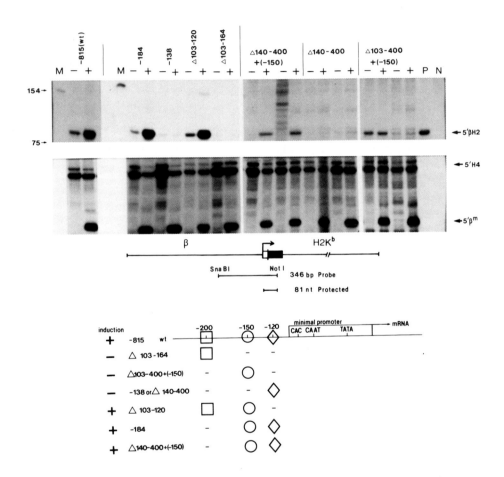

Fig. 2

Fig. 2:

S1 nuclease protection analysis of β-globin promoter constructs in MEL cells.

Upper panel: S1 nuclease protection analysis of mRNA isolated from MEL cells, stably transfected with different β-globin gene promoter deletions, coupled to the H2K gene (Antoniou et al., 1988). Numbers above the lanes indicate the various promoter deletions, - and + indicate before and after induction of the MEL cells, M are makers, P and N are positive and negative controls for β-globin-H2K mRNA. 5'-bH2 is the signal for the hybrid RNA, 5'-H4 and 5'-bm are controls for loading (histone H4mRNA) and induction (mouse bmajor mRNA) respectively. The input probe and the fragment protected from S1 nuclease digestion by the 5' end of the mRNA are shown in the lower part of this panel.

Lower panel: Summary of the S1 nuclease protection data. Open squares, circles and diamonds indicate the presence of the -200, -150 or -120 areas in combination with the minimal promoter (CAC, CAAT, TATA) in the b-H2K hybrid gene. -815 wt is the normal β-globin gene promoter extending 815bp 5' of the transcription initiation site. This gene is cloned on the plasmid pTM as described (Antoniou et al., 1988). -103-164 is an internal deletion between position 103 and 164 containing an 8bp ClaI linker. -103-400+(150) is a similar deletion with the -150 region re-inserted as an oligonucleotide. -138, 140-400, 103-120, -184 and 140-400+(150) are similar constructs with different co-ordinates as indicated by the numbers. It is worth noting that the differences in the distance between the CAC box and the -150 region is a whole number of helical turns in the wild-type (wt) construct and the mutants 103-140+(150) and 103-120, while a half turn is introduced in the construct 140-400+(150).

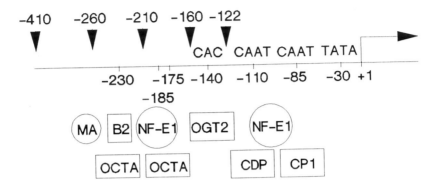

Fig. 3:

    Schematic representation of the $^A\gamma$-globin gene promoter.

    The symbols indicate the various protein factors NF-E1, CP1, etc. (see text). The lower co-ordinates indicate the position of the binding sites of these factors in the promoter.   The co-ordinates above the line show the deletions tested by Catala et al. (1988).   CAC, CAAT and TATA indicate the non tissue-specific motifs in the minimal promoter.   The arrow at +1 indicates the start of γ-globin transcription.

in the expression of the γ-globin gene (Ottolenghi, 1988; Orkin, 1988). Interestingly, the group of Ottolenghi has shown that the single base change in the HPFH results in the creation of another NF-E1 binding site (Mantovani et al., 1988). Together with the deletion experiments (Catala et al., 1988), this suggests that NF-E1 plays a stimulatory role in this part of the promoter. The third and last stimulatory region detectable by deletions binds a newly identified factor (B2 at -233, Fig. 3) present in K562 and MEL, but not HeLa, cell extracts (Catala et al., 1988). The region upstream of -260 has not been characterized in great detail; so far only one factor (MA, Fig. 3) has been shown to bind in this region (Mantovani et al., 1988).

In summary, we have identified a number of nuclear protein factors capable of binding to the regulatory regions flanking the human γ- and β-globin genes and in at least one case (NF-E1 in the β-globin promoter), it appears that protein-protein interactions are required for an induction of transcription. It is, however, not clear how any of the factors are involved in specific events at the molecular level. With the exception of some quantitative differences between the factors, we have not yet identified any clearly stage-specific differences in terms of protein factors. Although it is possible that this could be achieved through specific interactions with the DCR (as yet not analyzed in detail), this is not very likely, since the stage specific differences in expression of the γ- and β-globin genes are largely maintained, even in the absence of the DCR. It is more likely that yet other interacting factors or protein modifications are involved, but that the characterization of these will have to await further purification and cloning of some of the factors, in particular, NF-E1.

ACKNOWLEDGEMENTS

We are grateful to Cora O'Carroll for typing this manuscript. L.W. was supported by MRC (Canada), F.C. was supported by CNRS (France). This work is supported by MRC (UK).

REFERENCES

Anagnou N, Karlsson S, Moulton A, Keller G, Nienhuis A
    (1986). Embo J 5:121-126.
Antoniou M, deBoer E, Habets G, Grosveld F (1988). Embo J
    7:377-384.
Antonarakis S, Philips J, Mallonee R, Kazazian H, Fearon
    E, Weber P, Kronenburg H, Ullrich A, Meyer D (1983).
    Proc Natl Acad Sci USA 80:8615-8619.
Baron M, Maniatis T (1986). Cell 46:591-602.
Behringer RR, Hammer RE, Brinster RL, Palmiter RD, Townes
    TM (1987). Proc Natl Acad Sci USA 84:7056-7060.
Bodine D, Ley T (1987). Embo J 7:2997-3004.
Catala F, deBoer E, Habets G, Grosveld F (1988).
    Submitted.
Charnay P, Treisman R, Mellon P, Maniatis T and Axel R
    (1984). Cell 38:251-263.
Chodosh C, Baldwin A, Canthew R, Sharp P (1988). Cell
    53:11-24.
Collins F, Weissman S (1984). Prog Nucl Acid Res Mol Biol
    32:315-462.
Cowie A, Myers R (1988). Mol Cell Biol 8:3122-3128.
Davidson I, Xiao J, Rosales R, Staub A, Chambon P (1988).
    Cell 54:931-942.
deBoer E, Antoniou M, Mignotte V, Wall L, Grosveld F
    (1988). Embo J 7:4203-4212.
Dierks P, van Ooyen A, Cochran M, Dobkin C, Reiser J,
    Weissmann C (1983). Cell 32:695-706.
Evans T, Reitman M, Felsenfeld G (1988). Proc Natl Acad
    Sci USA 85:5976-5980.
Fletcher C, Heintz N, Roeder R (1987). Cell 51:773-782.
Forrester W, Takegawa S, Papayannopoulou T,
    Stamatoyannopoulos G, Groudine M (1987). Nucl Acids
    Res 15:10159-10177.
Grosveld G, Rosenthal A, Flavell RA (1982). Nucl Acids Res
    10:4951-4971.
Grosveld F, Blom van Assendelft G, Greaves D, Kollias G
    (1987). Cell 51:21-31.
Groudine M, Kowhi-Shigematsu T, Gelinas R,
    Stamatoyannopoulos G, Papayannopoulos P (1983). Proc
    Natl Acad Sci USA 80:1751-1755.
Jones K, Kadonega J, Rosenfeld P, Kelly T, Fjian R (1987).
    Cell 48:79-89.
Kioussis D, Wilson F, Khazaie K, Grosveld FG (1985). Embo
    J 4:927-931.

Kollias G, Hurst J, deBoer E, Grosveld F (1987). Nucl Acids Res 15:5739-5747.

Kollias G, Wrighton N, Hurst J, Grosveld F (1986). Cell 46:89-94.

Mantovani R, Malgretti N, Nicolis S, Giglioni B, Comi P, Capellini N, Bertero M, Caligaris-Cappio F, Ottolenghi S (1988). Nucl Acids Res 16:4299-4313.

Mantovani R, Malgretti N, Nicolis S, Ronchi A, Giglioni B, Ottolenghi S (1988). Nucl Acids Res 16:7783-7797.

Myers R, Tilly K, Maniatis T (1986). Science 232:613-618.

Orkin S, Antonarakis S, Kazazian H (1984). J Biol Chem 259:8879-8881.

Orkin S, Kazazian H, Antonarakis S, Goff S, Brehm C, Sexton J, Weber P, Giardina P (1982). Nature 296:627-631.

Orkin S, Sexton J, Cheng T, Goff S, Giardina P, Lee J, Kazazian H (1988). Nucl Acids Res 11:4727-4734.

Poncz M, Henthorn P, Stoeckert C, Surrey S (1988). Oxford Surveys on Eukaryotic Genes. In press.

Rutherford T, Nienhuis A (1987). Mol Cell Biol 7.398-402.

Schule R, Muller M, Otsuka-Murakami H, Renkawitz R (1988). Nature 332:87-90.

Superti-Furga G, Berberis A, Schaffner G, Busslinger M (1988). Embo J 10:3099-3108.

Townes T, Lingrel J, Chen H, Brinster R, Palmiter R (1985). Embo J 4:1715-1723.

Treisman R, Orkin S, Maniatis T (1983). Nature 302:591-596.

Trudel M, Magram J, Bruckner L, Costantini F (1987). Mol Cell Biol 7:4024-4029.

Tuan P, Solomon W, Qiliang L, Irving M (1985). Proc Natl Acad Sci USA 22:6384-6388.

Wall L, deBoer E, Grosveld F (1988). Genes and Develop 2:1089-1100.

Wright S, Rosenthal A, Flavell RA, Grosveld FG (1984). Cell 38:265-273.

Wrighton N, Grosveld F (1988). Molec Cell Biol 8:130-137.

Xiao J, Davidson I, Macchi M, Rosales R, Vigneron M, Staub A, Chambon P (1987). Genes Dev 1:794-807.

Hemoglobin Switching, Part A: Transcriptional Regulation, pages 15–35

REPLICATION AND ACTIVATION OF THE HUMAN β-GLOBIN GENE DOMAIN

Mark Groudine, William C. Forrester, Ulrike Novak, and Elliot Epner

Fred Hutchinson Cancer Research Center and the University of Washington School of Medicine Seattle, WA 98104

To study the molecular mechanisms involved in hematopoetic differentiation, our laboratory has focused on changes occurring at the level of chromatin, using the globin gene loci in normal and transformed erythroid cells as model systems. The human β-globin gene locus consists of six linked genes in the 5' to 3' order ε-Gγ-Aγ-ψβ-δ-β spanning approximately 50 kilobases (kb) of DNA on the short arm of chromosome 11 (1). In human fetal liver, adult erythroid cells, and erythroleukemic cell lines, different of the globin genes are transcribed, yet the entire β-like globin gene locus (including sequences flanking the structural genes) is preferentially sensitive to DNase I digestion (2). Although the exact boundaries of the DNase I sensitive domain have not been established, sequences approximately 45 kb 5' to the ε-globin gene adjacent to the (γδβ)° thalessemia breakpoint are DNase I resistant (3). In contrast, sequences at the 3' breakpoint of patients with the deletional form of hereditary persistence of fetal hemoglobin (HPFH) over 100 kb 3' to the β-globin gene (4) are DNase I sensitive (5). It is not known whether the 3' HPFH breakpoint region is part of the β-globin gene domain or part of a separate locus. In addition to the domain of overall DNase I sensitivity, two classes DNase I hypersensitive sites have been described. The first are erythroid and developmentally specific and are located at the 5' ends of the subset of globin genes active in a particular tissue or cell line (6). A second class of DNase I hypersensitive sites 6-18 kb 5' to the ε and 20 kb 3' to the β-globin genes are erythroid specific and developmentally stable, and are present regardless of which globin genes are active (2,7,8). Whether these sites form the boundaries of the DNase I sensitive domain has not been established.

In addition to their preferential sensitivity to DNase I, active genes replicate in early S phase (9). DNA replication has also been linked to the propagation of the active DNase I sensitive chromatin configuration (10), and has been thought to be important in the activation of new patterns of gene expression (11). In this regard, it has been proposed that loops or domains of active chromatin may correspond to a replication domain as defined by one or several simultaneously active origins of replication (12). A model which invokes changes in the pattern of origins of DNA replication in the in multigene families as a mechanism of transcriptional activation was proposed over 10 years ago (11,13), but little experimental evidence has accumulated to support this hypothesis. In an attempt to understand the relationships among DNA replication, chromatin structure, and gene activity, we have recently studied the timing of DNA replication in the human β-globin gene locus, and have begun to analyze the contribution of the developmentally stable hypersensitive sites to the formation of the globin domain.

TIMING OF REPLICATION IN THE β-GLOBIN DOMAIN

We first sought to examine the relationship between domains of DNA replication and active chromatin structure using the human β-globin gene locus as a model system. K562 erythroleukemia cells express a subset of the globin genes i.e. the embryonic ϵ- and ξ-globin genes, the fetal γ-globin genes, and the adult α- but not β-globin genes (2,6). Even though only a subset of the β-globin gene locus is expressed, the entire locus including adjacent flanking sequences is DNase I sensitive. Our first approach to map the domain of DNA replication in K562 cells was to determine the temporal ordering of DNA replication within the S phase, as has been done with the murine globin and immunoglobulin genes (14,15,16). The basic experimental design involves a short incubation of exponential cells with bromodeoxyuridine (BrdUrd) to density label newly synthesized DNA of exponentially growing cells. Cells are then separated into cell cycle specific fractions of early, middle, and late S phase cells by centrifugal elutriation (14,15). After DNA isolation and either restriction endonuclease digestion or random shearing, newly replicated BrdUrd containing DNA is isolated by density gradient centrifugation, and equal amounts from different elutriated fractions are Southern or slot blotted. These filters are then hybridized with globin or nonglobin probes.

In the experiment shown in Figure 1, exponentially growing K562 were labelled with BrdUrd for one hour and elutriated, and the cell fractions obtained were analyzed by flow microfluoremetry (FMF) using propidium iodide staining to determine DNA content and position in the cell cycle. Approximately 10 fractions corresponding in DNA content from 2-4 N were obtained. Nuclear DNA was isolated, digested with EcoRI, and BrdUrd-DNA purified by sodium iodide gradient centrifugation. Equal amounts of BrdUrd-DNA from 10 fractions corresponding to DNA synthesized during early, middle, and late S phase were electrophoresed, transferred to nylon filters, and hybridized with a variety of globin and non-globin probes. Figure 1B shows the results of hybridization to probes spanning over 120 kb of the β-globin gene locus. Virtually the entire β-globin gene locus replicates simultaneously in early S phase. The ε-globin gene which is actively transcribed in these cells replicates early in S phase, as does the β-globin gene which is not transcribed in K562 cells. By densitometry (not shown), over 80% of hybridization is in the first four fractions containing BrdUrd DNA synthesized during early S phase. Sequences from regions flanking the globin gene locus also replicate during this time. These include a 3.3 EcoRI fragment 20 kb 5' to the ε- globin gene and a 2.5 kb Sph fragment which hybridizes to a 4.1 kb EcoRI fragment about 45 kb 5' to the ε-globin gene. This fragment was cloned from the breakpoint of a patient with (γδβ)° thalessemia and resides in a region of chromatin which is heavily methylated and DNase I resistant in erythroid cells (3). In addition, probes from the 3' breakpoints of a patient with the Indian (17) and Black (4) hereditary persistence of fetal hemoglobin (HPFH) located 25 and over 100 kb 3' to the β-globin gene respectively, also hybridize to early replicating BrdUrd-DNA.

In contrast to the early replicating nature of the majority of the β-globin gene cluster, two probes within this region detect DNA sequences whose time of replication differed markedly from adjacent sequences. A γ-globin gene IVS-2 probe which hybridizes to two different EcoRI fragments (2.7 and 7.2 kb) containing both the Aγ- and the Gγ globin genes respectively revealed that the Aγ-gene (2.7 kb) replicates slightly later than the Gγ-gene (7.2 kb) although these genes are only 4 kb apart. The use of two probes located between the γ-globin genes confirms and extends this observation; a 0.7 kb EcoRI fragment is early replicating while the adjacent 1.6 kb EcoRI fragment replicates much later and seemingly throughout S (Figure 1B).

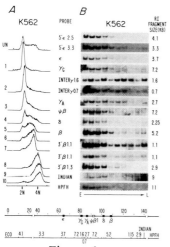

Figure 1

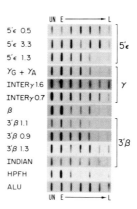

Figure 2

Figure 1. A) Distribution pattern of cellular DNA content of K562 cells growing exponentially in the presence of BrdUrd prior to fractionation according to size by centrifugal elutriation. An aliquot from each fraction was prepared, stained with propidium iodide and analyzed by flow cytometry as described (22). The top panel shows the unfractionated population and subsequent fractions correspond to early, middle, and late S phase cells.

B) Blot hybridization of newly synthesized BrdUrd-DNA from elutriated fractions of K562 cells to probes spanning the human β-like globin gene locus. 10 ugs of BrdUrd-DNA were cleaved with EcoRI, Southern blotted and hybridized as described in Materials and Methods. Each lane represents DNA from an elutriated fraction from early S (left) to late S phase (right). The location of each probe used and the EcoRI fragment hybridized in kb are shown. The map is taken from reference 24. Two exposures of the $3'\beta$ 1.1 kb EcoRI fragment are shown to emphasize its different pattern of replication. The signal seen in lane 1 using the inter-γ 0.7 kb probe has a superimposed background artefact. Probes used are described in references 10, 11 and 16 except for the following: $5'\epsilon$ 2.5-a 2.5 kb Sph fragment (3), $5'\epsilon$ 3.3-a 3.3 kb EcoRI fragment (see map in ref. 29), $3'\beta$ 1.3-a 1.3 kb EcoRI-BglII fragment (see map in ref.29), Indian-a 3' breakpoint 0.7 kb Hinf-BglII fragment (17), and HPFH-a 3' breakpoint 2.4 kb Bam-EcoRI fragment (4).

Figure 2. Slot blot hybridization of sheared BrdUrd-DNA to β-like globin gene probes. Slot blots were prepared using DNA from elutriated cell fractions, hybridized to β-like globin gene probes, and filters rehybridized as detailed in Materials and Methods. Unfractionated BrdUrd-DNA was loaded on the first slot with early to late S BrdUrd-DNA in individual slots from left to right as shown. The probes used are indicated on the left and are arranged in 5' to 3' order (see map in Figure 2). Slots were hybridized to a repetitive sequence Alu-containing probe to control for equal loading of DNA. Probes were identical to those used in Figure 1B and references 2,5, and 8 except for $5'\epsilon$ 0.5-a 0.5 kb BglII-EcoRI fragment and $3'\beta$ 0.9-a 0.9 kb EcoRI-Hind III fragment (see maps in ref. 29).

A similar but less striking result was observed in a region located about 20 kb 3' to the β-globin gene. A 1.1 kb EcoRI fragment replicates significantly later than a 2.9 kb EcoRI fragment located 5 kb 3' to it, although not as late nor as extended throughout S phase as the inter-γ 1.6 EcoRI fragment. Densitometry confirms that while >95% of hybridization to the 2.9 kb EcoRI fragment is found in the first four fractions, only 75% of hybridization to the 1.1 kb probe is to the first four early replicating fractions; the remaining 25% of hybridization is found in the remaining fractions containing BrdUrd-DNA synthesized during middle and late S. It is of potential interest that an erythroid specific developmentally stable DNase I hypersensitive site is located in this region (7,8).

These results were confirmed independently by hybridization to slot blots containing sheared BrdUrd-DNA from elutriated fractions of K562 cells labelled with BrdUrd for two hours (Figure 2). The β-globin gene is early replicating while the γ region replicates later in S phase. A γ IVS-2 probe hybridizes to BrdUrd-DNA that was later replicating than the β-globin gene; this is confirmed by densitometry (not shown) which shows that 65% of the β-globin hybridization is to the early (first) fraction, whereas 45% of the γ-globin hybridization is found in the early fraction. The inter-γ 1.6 and 0.7 EcoRI fragment probes hybridize to DNA sequences that replicate slightly later those hybridizing to the γ IVS-2 probe. However in contrast to the results obtained by Southern analysis of EcoRI digested DNA (Fig. 1B), slot blot assay of sheared DNA indicates that the 1.6 and 0.7 kb EcoRI fragments hybridize to DNA sequences that replicate at similar times (see below). These same slots were also hybridized with probes representing sequences located about 20 kb 5' to the ε and 20 kb 3' to the β-globin genes. These probes are adjacent to the 5' ε and 3' β DNase I hypersensitive sites that may determine the putative erythroid DNase I domain boundaries (7,8). A 1.3 kb probe corresponding to DNA sequences about 10 kb 5' to the ε globin gene hybridizes to early replicating sequences (70% of hybridization to fractions 1 and 2) while the 3.3 kb EcoRI fragment located about 20 kb 5' to ε replicates significantly later in S phase (50% of hybridization to fractions 1 and 2). A 0.5 kb EcoRI-BglII fragment located at the 5' end of the 3.3 kb EcoRI fragment hybridized to BrdUrd-DNA that is even later replicating (30% of hybridization to fractions 1 and 2) suggesting that a replication domain boundary is located just 5' to the 3.3 EcoRI fragment. This region contains highly repetitive DNA sequences for which single copy probes are not presently available (18). A replication boundary region in the region 20 kb 3' to β is also apparent. The 1.1 kb EcoRI fragment is later replicating than the β-globin gene while probes located about 5 kb 3' (0.9 EcoRI-HindIII and 1.2 EcoRI-BglII) are even later replicating. Since this region is also filled with repetitive sequences the exact location of this late replicating region cannot be mapped with certainty. However a probe for the 3' breakpoint of a patient with the Indian HPFH phenotype (3' IH) located about 10 kb further 3' (17) also hybridizes to slots containing predominantly later replicating BrdUrd-DNA while the HPFH breakpoint probe located over 100 kb further 3' again hybridized to early replicating BrdUrd DNA.

Apparent differences in the replication pattern of specific DNA sequences obtained with identical probes using slot versus Southern analysis can be explained by different methods of DNA preparation. For example, while the 3.3 EcoRI fragment is early replicating by Southern analysis, its later replication behavior on slot blots of sheared DNA reflects the presence of later replicating sequences just 5' to it. The later replication behavior of sequences homologous to the 0.5 kb probe at the 5' end of this fragment and the earlier replication of sequences detected by the 1.3 kb 3' probe support this. In addition, the replication patterns of the inter-γ 1.6 and 0.7 EcoRI fragments appear quite different by Southern blotting, but are similar on slot blots of sheared DNA, due to the physical proximity of the two fragments and the relatively large size of the sheared slotted DNA (10 kb).

A variety of probes for genes presumably unexpressed and DNase I resistant in these cells were hybridized to the Southern blot shown in Figure 1B (data not shown). The immunoglobulin $C\mu$ gene probe (19) and the $\alpha2(I)$ collagen probe (5) hybridized to fragments replicating in middle S phase, while the J_H (19) immunoglobulin probe detects sequences that replicate slightly later in S phase.

The Burkitt's lymphoma B cell line Manca was similarly labelled with BrdUrd for 2 hours, elutriated, and BrdUrd-DNA isolated and analyzed. The FMF of the cell fractions obtained is shown in Figure 3A. Hybridization of Southern blots containing BrdUrd-DNA to β-globin locus probes shows that the entire locus replicates in mid to late S phase in Manca cells where they are not expressed and DNase I resistant (20). However the α- (not shown) and ξ-globin genes gave hybridization patterns consistent with an early replication pattern even though these genes are DNase I resistant and not expressed. Probes for the c-myc and immunoglobulin J_H genes hybridize to early replicating DNA, consistent with these genes being expressed and DNase I sensitive in Manca cells. As shown previously both the germline and rearranged c-myc alleles are transcribed in Manca cells (21), and both replicate early in S phase (Figure 3B).

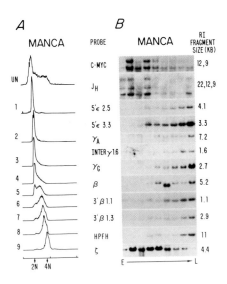

Figure 3

Figure 3. A) Distribution pattern of cellular DNA content of exponentially growing Manca cells fractionated according to size by centrifugal elutriation.
B) Blot hybridization of BrdUrd-DNA from elutriated Manca cells. Filters prepared as in Figure 2 were hybridized to probes as in Figures 2 and 4. The increased hybridization seen in the sixth lane of the hybridization to the β-globin probe and in the last lane for many probes is due to plasmid contamination and overloading of DNA respectively. Probes used were identical to those in Figures 1B, 2, and 4 and reference 21 for the c-myc probe.

MODELS TO EXPLAIN THE REPLICATION BEHAVIOR OF THE β-GLOBIN DOMAIN

Our analysis of the temporal ordering of DNA replication in the human β-globin gene locus extended over a contiguous region greater than 120 kb, and included sequences located over 100 kb 3' to the β-globin gene. In the lymphoid cell line Manca, in which the β-like globin genes are transcriptionally silent, the entire locus replicates in mid to late S phase. In contrast the majority of the β globin domain replicates synchronously in early S phase in K562 cells, including genes that are transcribed (e.g. ε-globin), not transcribed (β-globin), and flanking regions that are both DNase I sensitive and insensitive (summarized in Figure 4). Thus the domain of early replication for the human β-globin gene locus in K562 cells seems to extend beyond the domain of DNase I sensitivity, although the exact boundaries of these domains have not been precisely defined. In addition we have investigated the timing of replication of a number of other linked and unlinked genes, some of which are transcriptionally active and DNase I sensitive and others which are not transcribed and are DNase I resistant. Genes which are active all replicate early in S phase in cells where they are expressed. However inactive genes replicate in a heterogenous fashion; in some cases early in S phase (α- and ξ- globin genes in Manca cells) while in other cases in middle or late S phase. Thus early replication seems to be necessary but not sufficient for transcriptional activity. This is in agreement with the results of others (see ref. 3 for review).

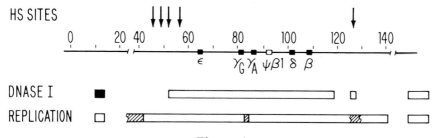

Figure 4

Figure 4. Summary of the replication behavior, DNase I sensitivity and location of DNase I hypersensitive sites in the human β-globin gene locus. Black areas represent DNase I resistant chromatin and white areas DNase I sensitive chromatin. Hatched regions represent regions of asynchronously replicating DNA (jagged edges represent approximate boundaries) and white boxes early replicating DNA sequences. Only the location of erythroid specific, developmentally stable DNase I hypersensitive sites are shown. See text for details.

Previous work using similar techniques demonstrated that the α- and β-globin genes were early replicating in murine erythroleukemia (MEL) cells, both before and after induction of erythroid differentiation by chemical inducers (14,15,22). The linked but unexpressed embryonic β-like globin gene was also early replicating (22). The inactive immunoglobulin heavy chain constant region genes were also studied in MEL cells (23). These genes replicated in early middle to late S phase in an order indicative of their linear array in the genome, consistent with the movement of a single replicon at approximately 2 kb/ minute. However analysis of the timing of replication of the heavy chain constant region genes in murine lymphoid cells expressing these genes revealed that the entire locus replicated simultaneously early in S phase, a result consistent with multiple origins being active (16). In preliminary experiments we have observed that the heavy chain constant region genes in K562 cells have a replication pattern similar to MEL cells (not shown). These observations suggest that in addition to the time of replication in S phase, the number of active origins might be important in establishing gene activity. However, over 120 kb of the β-globin locus as well as a probe over 100 kb 3' to β replicate simultaneously early in K562 cells and simultaneously late in Manca cells. These results suggest that more than one replicon is active in the β-globin locus in cells that express the globin genes as well as in cells in which the globin genes are silent, but that the time of activation of these multiple origins differs in the two cell types. Thus it will be of interest to determine the timing of replication of other multigene domains in various cell types.

The most surprising result of our analysis of the timing of DNA replication of the β-globin like gene locus in K562 cells is the identification of several areas which replicate unusually. The first, a 1.6 kb EcoRI fragment in the inter γ-globin gene region appeared to replicate homogenously throughout S phase while neighboring DNA sequences were early replicating. The second a 1.1 kb EcoRI fragment located about 20 kb 3' to the β -globin gene, replicated significantly later than adjacent sequences although not as homogenously throughout S as the γ -1.6 fragment. This region is also near the site of a 3' erythroid specific DNase I hypersensitive site and may mark the 3' boundary of the erythroid domain of DNase I sensitivity (7,8). Although the exact location and extent of this

3' replication boundary has not yet been mapped, slot blots analysis of sheared DNA suggests it is larger than the inter γ region. In addition we have also obtained evidence for a similar region just 5' to the DNase I hypersensitive sites 6-20 kb 5' to the ε-globin gene. This region is located just 5' to the 3.3 EcoRI fragment (see Figure 2) and appears similar to the 3'β region and larger in extent than the inter-γ region by its hybridization behavior on slot blots. We have recently observed the unusual replication behavior of these sequences in the human erythroleukemia cell line HEL (5) (E. Epner and W. Forrester, unpublished). Thus, this pattern of replication appears to be erythroid-specific, but not limited to K562 cells. Although the 5' and 3' areas of asynchronous replication appear larger than the inter-γ region, they do not define transition zones of early replicating DNase I sensitive chromatin from outlying late replicating and DNase I resistant chromatin, since Southern blot analysis reveals that sequences outside the 5' and 3' boundary regions are early replicating (Figure 4). These distant early replicating sequences include the DNase I resistant 2.5 kb Sph fragment about 45 kb 5' to ε-globin and the sequences homologous to the Indian and HPFH 3' deletion breakpoint probes, which are located 30 kb and over 100 kb 3' to β-globin, respectively.

The existence of islands of continously replicating DNA have not previously been described in eukaryotic chromosomes and seem difficult to reconcile with previous concepts of DNA replication. However, several models can explain these results. In the first, these regions represent potential termination sequences when two adjacent replication forks moving bidirectionally meet. Here, the rate of chain elongation might be slowed, resulting in areas that appear to replicate throughout or late in the S phase. Alternatively, the replication of this region might be delayed if decatenation preceded the completion of DNA synthesis (24). In our experiments differentiation between replication late or throughout S phase is difficult, because of the lack of purity of middle and late S elutriated fractions (see Figure 1). This colliding replication fork phenomenon has been observed in replicating SV40 molecules (24) where the termination points for DNA replication are located at the convergence of the two replication forks about 500 bps apart and 180 degrees from the origin of replication. While it remains possible that specific sequences define the sites of termination of replication forks (25), experiments using SV40 deletion mutants suggests that specific DNA termination sequences are not required (24). Thus, it will be interesting to analyze the replication timing of the β-globin locus in naturally occurring mutants which contain deletions in and adjacent to the inter-γ region (reviewed in 1).

Another possible explanation of our data is that these areas represent termination regions but that the replication fork moves unidirectionally, as has been documented for approximately 10% of the replication bubbles observed by autoradiography (26). This would imply that a replication origin is located adjacent to the termination region. A recent report (27) has intimated the proximity of replication origins and termination sequences in pea root cells. In preliminary experiments, we have observed sequences adjacent to the asynchronously replicating regions that replicate significantly earlier than the majority of DNA in the locus in the 3'β region in K562 and HEL cells. Diffferentiation between these models will require the determination of the directionality of fork movement.

It is also possible that these regions represent origins of replication. In this model, multiple abortive initiation events, due perhaps to an excess of an initiator-like protein (28), would result in continuous replication of these regions. If these sequences were replicating many times during a single S phase, one might expect both DNA strands to contain BrdUrd. However, if the time between reinitiations is greater that the length of the BrdUrd pulse, only one strand would contain BrdUrd. Thus, this model is directly testable.

The distance between these three regions of apparently continuously replicating DNA in the human β-globin gene cluster (about 45-50 kb apart) is in agreement with estimates of potential inter-origin distances by others (13,16,26). If these regions do represent termination sites and the replication forks are moving bidirectionally, origins of DNA replication should be located approximately midway between adjacent islands, near the β-globin and ϵ-globin genes. Although the very early replication of the β-globin gene region by slot-blot analysis supports this idea, we have not yet detected a similar region near the ϵ-globin gene. Thus, it will be important to map precisely the location of origins and termination points for DNA replication in the β-globin gene locus and to determine if changes in the patterns of origins used can influence the developmental regulation of these loci.

ACTIVATION OF THE β-GLOBIN DOMAIN

Attempts to understand the mechanisms which regulate the activation of the β-globin genes have been facilitated by naturally occurring mutant globin alleles which exhibit abnormal patterns of regulation. For example, in a specific subset of the $(\gamma\delta\beta)^{\circ}$ thalassemia class of mutations, deletions of the 5' end of the β-globin cluster are associated with the inactivation and DNase I resistant chromatin structure of all of the remaining globin genes located in cis (3). Although the β-globin genes which reside on the deleted chromosome are not expressed, they function normally when assayed in heterologous transcription systems (3). The finding that the DNA juxtaposed into the globin locus as a result of these deletions is DNase I resistant in its native chromosomal location led to the hypothesis that elements within this region exert a negative effect on the remaining globin genes (3). However, since these deletions remove the region containing the developmentally stable hypersensitive sites 5' to the ϵ-globin gene, we postulated that the failure to activate the globin genes in these thalassemias was due to the removal of the positive elements necessary for the activation of the globin domain early in erythroid differentiation (2).

Recent experiments using somatic cell hybrids (8) and transgenic mice (29) have provided evidence that the developmentally stable hypersensitive sites may be important in the activation of the β-globin locus. For example, in somatic cell hybrids resulting from the fusion of human non-erythroid cells and murine erythroleukemia (MEL) cells, the human adult β-globin gene, but neither the embryonic nor fetal globin genes, is activated transcriptionally (8). However, the DNase I resistant β-globin locus characteristic of the parental non-erythroid human cells is reorganized into a DNase I sensitive domain over an approximately 80 kb region, including the formation of the developmentally stable hypersensitive sites 50 kb 5' and, to a lesser extent, 20 kb 3' to the activated adult β-globin gene. These results are consistent with the hypothesis that the 5', and possibly 3', developmentally stable hypersensitive sites may be important for the activation of the β-globin chromatin domain. Thus, within the context of this model, the developmentally stable hypersensitive sites were termed "Locus Activation Regions" or LARs (8). A corollary of this hypothesis is that the high level expression of any of the β-like globin genes may require a restructuring of chromatin that encompasses the entire β-globin chromatin domain. This would predict that attempts at reproducing normal globin gene regulation will depend on the formation of this domain. Direct support for this model was obtained by Grosveld and colleagues, who demonstrated that inclusion of approximately 40 kb of sequence containing the 5' and 3' LARs in β-globin constructs introduced into transgenic mice resulted in high level, copy number dependent and position-independent expression of the β-globin gene (29). This is in marked contrast to the low level, variable, position-dependent and copy number independent expression of the β-globin transgenes which lack the LAR (30,31,32). Similarly, previous studies in which the β-globin gene was transfected into MEL cells demonstrated that although the gene was properly regulated with respect to induction, the levels of expression were considerably lower than the endogenous mouse globin genes (33).

To further understand the relationship between the LARs and the position independent expression of the linked β-globin gene, we have constructed cassettes containing DNA sequences corresponding to all four or subsets of the 5' developmentally stable hypersensitive sites. We chose to omit the site 3' to the β-globin gene from our cassettes, principally because the DNA sequences corresponding to the location of this hypersensitive site is present in the previously described (γδβ)° thalessemias deletions yet no activation occurs. Furthermore, this hypersensitive site was very weak and only variably reformed in our somatic cell fusion studies (8). In addition, Grosveld et. al. (29) reported the high-level expression of a human β-globin transgene containing the 5'ε region but from which the 3'β region had deleted spontaneously.

Experimental strategy

Figure 5 shows the native positions of the 5' hypersensitive sites with respect to the ε-globin gene, and the restriction fragments that were ligated together to construct the 8 kb mini-LAR (mLAR), 2.5 kb micro-LAR (μLAR) and other LAR derivatives. The human β-globin gene, including approximately 600 bp of 5' flanking sequence and 1.7 kb 3' to the poly A addition site, was inserted 3' to the various LAR constructs. This β-globin gene contains all previously described regulatory sequences within and 3' to the gene, including the putative stage-specific β-globin gene enhancer and elements responsible for increased transcription after induction (34-37). To assay the possible regulatory effects of the LAR sequences, we introduced the human β-globin gene with or without the various LAR cassettes in cis into MEL cells by electroporation. The SV2*Neo plasmid (38) was co-electroporated in all experiments, and G-418 resistant clones were isolated. The amount of human and mouse adult β-globin RNA in the induced state was quantitated by RNase protection, and the copy numbers of the human β-globin gene were determined for each clone by Southern blotting. The ratio of induced human β-globin RNA per human β-globin gene was normalized to the amount of endogenous mouse β-major (β^{maj}) globin RNA in each clone, which is an internal control for the induced levels of globin RNA in these cells. The number of mouse globin genes in each of the aneuploid MEL clones is unknown; the ratio of per copy expression of the human β-globin genes to mouse β^{maj} RNA provides a conservative estimate of the relative activity of the human to mouse genes since it is equivalent to expressing human to mouse globin gene activity assuming one copy of the mouse globin gene per MEL cell.

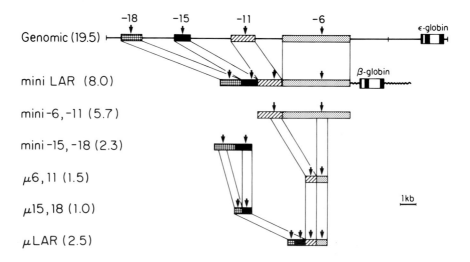

Figure 5

Figure 5. The positions of the developmentally stable hypersensitive sites 5'
to the ε-globin gene. The sites are described by their distance upstream
from the epsilon globin transcription initiation site, in kb. Relatively large
restriction fragments containing the hypersensitive sites were ligated
together and placed 5' to the β-globin gene as shown in the second line to
make the mini LAR. The 8 kb mini LAR was subdivided as shown and the
LAR fragments inserted 5' to the β-globin gene as described for the mini-
LAR. The length of the LAR derivatives, in kb, is given in parentheses.

---

mLAR and μLAR confer high level expression

The initial question we addressed was whether high level expression
of a human β-globin gene could be achieved in MEL cells by linkage in cis
to a LAR cassette from which the 3'β hypersensitive site region and
sequences between the four 5'ε hypersensitive sites were deleted. The
results of this analysis are shown in Table I. In clones containing the β-
globin gene without LAR sequences, the amount of human β-globin RNA
per gene copy ranges from 3% to 16%, with an average of 10%, of the
amount of mouse β^{maj} globin RNA in induced MEL cells. This is in
agreement with previous reports (33,39). In contrast, all clones containing
the human β-globin gene linked to the LAR and subsets of LAR sequences
express high levels of human β-globin RNA following induction. On a per
gene basis, clones containing the mLAR/β-globin cassette contain 20 to
120% (average=80%) the amount of human β-globin RNA compared to
mouse β^{maj} globin RNA. Similarly, clones into which the μLAR/β-globin
constructs have been introduced produce between 50 and 250%
(average=108%) the amount of human compared to mouse β-globin RNA,
when normalized per human globin gene copy. In two clones containing a
cassette in which the μLAR is linked to the human β-globin gene in the
reverse genomic orientation (μRAL), the amount of human β-globin RNA
per gene is 50 and 220% (average=135%) that of mouse β^{maj} globin RNA.
These results demonstrate that the mLAR and μLAR cassettes can increase
the expression of a linked β-globin gene to levels that on average are similar
to the expression of the endogenous mouse β-globin genes in MEL cells, and
that the effect of the LAR is orientation independent.

Cell line	Copy #	hβ RNA/gene copy / mβmaj RNA
β 1	1	0.1
β 2	>30	<0.1
β 3	1	0.03
β 4	1	0.16
mLAR 1	5	0.5
mLAR 2	5	1.1
mLAR 3	1	1.1
mLAR 4	2	0.3
mLAR 5	1	1.2
mLAR 6	6	1.2
mLAR 7	3	0.2
μLAR 1	2	1.8
μLAR 2	2	0.5
μLAR 3	2	2.5
μLAR 4	1	0.6
μRAL 1	2	0.5
μRAL 2	2	2.2
m15,18 1	4	0.4
m15,18 2	3	0.4
m15,18 3	6	0.5
μ6,11 1	2	0.3
μ6,11 2	5	0.2
μ6,11 3	2	0.4

Table 1. Quantitative comparison of human β-globin:mouse β-globin expression levels in MEL cells. Cell lines, listed on left, which contain the human β-globin genes were analyzed for the number of integrated human β-globin genes by Southern blotting. Ratios of human β RNA to mouse β^{maj} RNA in cells induced with 2% DMSO were determined by counting the excised gel region corresponding to the protected globin mRNA fragment, and correcting the amount of human β-globin RNA to the number of integrated genes.

The experiments described above indicate that elements within the mLAR and μLAR lead to high level expression of the human β-globin gene introduced into MEL cells. To determine if hypersensitive sites are formed at their natural positions on the LAR derivative after introduction into these cell lines, we analyzed the chromatin structure of the μLAR and mLAR/human β-globin gene cassettes in several of the MEL clones. Our results reveal that all four 5'ε hypersensitive sites are formed in the stably integrated mLAR and μLAR β-globin gene cassettes in a sequence-dependent manner and independent of orientation (Figure 6). In addition, these sites are reformed both prior to and after induction of the MEL cells.

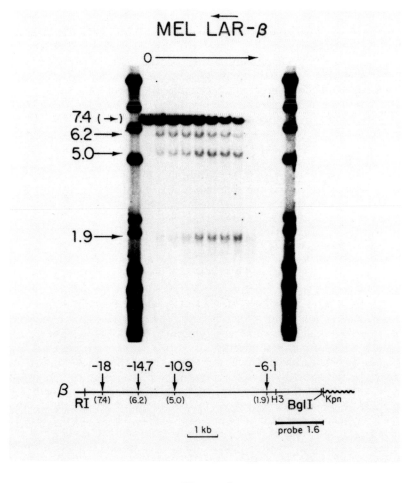

Figure 6

Figure 6. Chromatin structure of the mini-LAR (in the opposite genomic orientation to the linked β-globin gene in a single copy cell line before induction. DNA from DNase I treated nuclei was digested with EcoRI and BgII, electrophoresed, blotted and hybridized with the 1.6 kb EcoRI-HindIII probe. Positions of DNase I hypersensitive sites within the mini-LAR are shown in kb from the BgII site in parentheses beneath the original, genomic location of these hypersensitive sites relative to the ε-globin gene. Wavy line indicates flanking cellular DNA. Size markers are λ/HindIII and φX/HaeIII end-labeled restriction fragments.

Tissue specificity of the LAR effect

To address the question of tissue specificity of the LAR effect, the human β-globin gene described above with and without the mLAR was introduced into NIH 3T3 cells by electroporation. In the absence of the LAR, the β-globin gene is transcribed, but the levels are variable and do not correlate with the number of β-globin gene copies (not shown). This result is consistent with previous observations regarding the activity of β-globin genes introduced into non-erythroid cells by transfection (40-44). When the mLAR is linked in cis to the human β-globin gene in either the genomic or reverse genomic orientation (not shown) and introduced into 3T3 cells, no effect on the steady-state levels of human β-globin RNA is observed. In addition, there is no correlation between the quantity of human β-globin RNA and the number of LAR linked β-globin genes.

If the hypothesis that the LAR exerts its effect through chromatin structure is correct, the failure of the LAR to influence β-globin gene expression in 3T3 cells might suggest that the usual pattern of the erythroid specific hypersensitive sites would not be formed in these cells. Therefore we assayed the chromatin structure of the LAR sequences in seven independently isolated mLAR/β-globin cassette containing 3T3 clones. Our results show that a hypersensitive site is reformed in the mLAR sequences corresponding to the previously described erythroid specific hypersensitive site located -10.9 kb to the ε-globin gene (not shown). This hypersensitive site was formed equally well clones containing the LAR in either the genomic or reverse genomic orientation with respect to the human β-globin gene. The formation of this hypersensitive site in the 3T3 clones is surprising, since none of the 5'ε hypersensitive sites were found in our previous analyses of the chromatin structure of non-erythroid cells. This unexpected result may be analogous to the finding that the transfected, but not the endogenous β-globin gene is expressed in 3T3 and other non-erythroid cells, and may reflect differences between the regulation of endogenous sequences and those that integrate into the genome after transfection and transient activity.

LAR Reductions

To further define the elements within the LAR responsible for the high level, tissue specific expression of the linked β-globin gene, cassettes containing the human β-globin gene and LAR sequences corresponding to the -6 and -11 hypersensitive sites or the -15 and -18 sites were introduced into MEL cells. As shown in table 1, human β-globin gene activity is several fold higher in MEL clones containing these cassettes compared to MEL clones containing the β-globin gene alone. However, neither the 6,11 nor the 15,18 cassettes confer as high expression as the complete mLAR or μLAR to the linked β-globin gene. In cells containing the mLAR 6,11 cassette, the levels of human β-globin RNA per gene copy ranged from 20 to 40% of mouse β^{maj} globin RNA (average = 30%), and in those containing the μLAR 15,18 cassette, the per gene copy expression of human β-globin gene is 40 to 80% that of mouse β^{maj} globin RNA (average = 43.5%). Analysis of the chromatin structure of the 6,11 and 15,18 LAR reductions (not shown) reveals that the appropriate hypersensitive sites are reformed in MEL cells, reinforcing the correlation between activity of the LAR elements and the formation of these alterations in chromatin structure.

IMPLICATIONS FOR GENE EXPRESSION

The results presented above show that the sequences retained after reduction of the 5'ϵ region from 20 kb to 2.5 kb (μLAR) can confer the orientation independent, high level expression of a cis-linked β-globin gene in stable MEL transformants. We have also shown that as little as 1500 bp of the μLAR (μ6,11) and a cassette containing the -15 and -18 hypersensitive site regions significantly increase β-globin gene expression in MEL cells. Interestingly, however, the two "half" LARs are not as active as the complete mLAR or μLAR. This result suggests that while the active elements within the LAR can function independently, these elements may work co-operatively or additively in influencing the level of expression of the linked β-globin gene. Distinguishing between the additive versus co-operative effects of these elements will require the construction of cassettes containing duplications of the 6,11 and 15,18 elements.

Our chromatin analyses revealed that all four DNase I hypersensitive sites are formed on the mLAR and μLAR before and after induction and independent of the orientation of the LAR sequences with respect to the β-globin gene. Similarly, hypersensitive sites are formed on constructs made from LAR subfragments containing sequences corresponding to the -6 and -11 or -15 and -18 sites. The effect of the LAR sequences on β-globin gene expression is tissue-specific: in 3T3 cells, LAR/β-globin gene cassettes are expressed at a low level, independent of copy-number, similar to the expression of the β-globin gene without LAR elements. In addition, only the -11 hypersensitive site is reformed on the LAR/β-globin gene cassette stably integrated in 3T3 cells.

The formation of the hypersensitive sites on the mLAR and derivative cassettes prior to induction of MEL cells suggests that while these sites may exert their effect by conferring an active chromatin structure to the linked gene independent of integration site, this active structure is not sufficient for the optimal expression of the linked β-globin gene. This result is consistent with our previous findings that the human β-globin domain is DNase I sensitive and the LAR hypersensitive sites are present in K562 cells and in the human/MEL hybrids prior to induction and overt transcription of the globin genes (2,5,8). We have referred to this active state of chromatin present in cells prior to transcription as a "preactivation" state (45,46,47). Within the context of this model, genes residing in the preactivation chromatin structure may have increased accessability to gene-specific (i.e. promoter, enhancer, etc.) factors dependent upon the induced state for their activity.

If the above model is correct, other non-globin genes which do not respond to MEL induction signals should be efficiently expressed in uninduced MEL cells when linked to the LAR. As a preliminary test of this prediction, we have determined the number of G418 resistant MEL colonies produced after introduction of cassettes consisting of the HSV Tk promoter and the neomycin resistance gene with and without the mLAR. After normalizing the number of G418 resistant colonies to a co-electroporated transformation control, the mLAR-Tk-Neo cassette produced approximately 35-50 times the number of G418 resistant colonies compared to the Tk-Neo genes alone (Forrester and Novak, unpublished observation). These results are consistent with the hypothesis that the LAR facilitates influences expression of genes in cis, in a manner that is not dependent on the induction associated events required for high level expression of the globin genes in MEL cells. These results are also compatible with, but certainly do not prove, the notion that the LAR effect is mediated through effect on gene expression is mediated through the establishment, maintenance, and propagation of an active chromatin structure. With respect to this model, the absence of LAR sequences may subject genes to a metastable chromatin structure, which may alternately permit or inhibit expression. This metastable reorganization of chromatin may, in part, explain the variable and low levels of the human β-globin gene without the full LAR in MEL cells and with or without the erythroid-specific LAR in 3T3 cells (see ref. 48 for review).

Sequence analysis of the μLAR constituent elements reveals that each of the four hypersensitive site regions contains enhancer core-like sequences sand significant stretches of alternating purines and pyrimidines (7). In addition, the μ6 hypersensitive site region contains an oligo T stretch, and the μ15 element contains three consensus NFE-1 (ERYF-1) binding sites (49,50). Thus, all elements of the μLAR contain sequences that may be involved in the generation of altered DNA structures and/or transcriptional regulation. Recently, Driscoll et. al. (51) have described a new deletional $(\gamma\delta\beta)^o$ thalassemia, in which sequences corresponding to the -6.1 hypersensitive site are retained 3' to the breakpoint region, and the -11,-15 and -18 regions are all deleted (unpublished results). This novel deletion suggests that sequences involved in the formation of the -6.1 hypersensitive site are not sufficient by themselves to influence the activation of the cis-linked β-globin locus. Consistent with this, we have found that the -6.1 hypersensitive site is not formed and the entire β-globin locus on the deleted chromosome is DNase I resistant in somatic cell hybrids resulting from the fusion of MEL cells and the patient's lymphoid cells (unpublished results). Thus, in an attempt to further localize the sequences within the μLAR responsible for activation of the cis-linked β-globin gene, we have focused on the sequences contained within the -11 element. Our preliminary results suggest that a cassette containing the μ11 element activates a linked β-globin gene as efficiently as the μ6,11 cassette (unpublished). The demonstration that small fragments containing LAR elements can effectively increase the level of a cis-linked β-globin gene has led to the construction of retroviral vectors that are capable of significantly increased β-globin gene expression in MEL cells (Novak et. al., in preparation).

REFERENCES

1. Collins, F.S. and Weissman, S.M. (1984) Prog. Nucleic Acid Res. Mol. Biol. 31:315-458.
2. Forrester, W.C., Thompson, C., Elder, J.T., and Groudine. M. (1986) Proc. Nat'l Acad. Sci. USA 83:1359-63.
3. Kioussis, D. Vanin, E., deLange, T., Flavell, R.A., and Grosveld, F.G. (1984) Nature 306:662-666.
4. Tuan, D., Feingold E., Newman, M., and Forget, B. (1983) Proc. Nat'l Acad. Sci. USA 80:6937-6941.
5. Elder, J.T., Forrester, W., and Groudine, M. Manuscript in preparation.
6. Groudine, M., Kohwi-Shigematsu, T., Gelinas, R., Stamtoyannopoulos, G., and Papayannopoulou, T. (1983) Proc. Nat'l. Acad. Sci. USA 80:7551-7555.
7. Tuan, D., Solomon, W., Li, Q., and London, I.M. (1985) Proc. Nat'l Acad. Sci. USA 82:6384-6388.
8. Forrester, W.C., Takegawa, S., Papayonnopoulou, T., Stamatoyannopolos, G., and Groudine, M. (1987) Nucleic Acids Res. 15:10159-10177.
9. Holmquist, G.P. (1987) Am. J. Hum. Genet. 40:151-173.
10. Weintraub, H. (1979) Nucleic Acid Res. 7:781-792.
11. Weintraub, H., Flint, S., Leffak, I., Groudine, M., and Grainger, R. (1977) Cold Spring Harbor Symp. Quant. Biol. 42:401-407.

12.   Conklin, K. and Groudine, M. (1984) in Razin, A. and Riggs,A.D.
       (eds): "DNA Methylation" New York: Springer pp. 293-351.
13.   Smithies, O. (1982) J. Cell Physio. Supp. 1:137-143.
14.   Epner, E., Rifkind, R., and Marks, P.A. (1981) Proc. Nat'l Acad. Sci.
       USA 78:3058-3062.
15.   Furst, A., Brown, E.H., Braunstein, J.D. and Schildkraut, C. (1981)
       Proc. Nat'l. Acad. Sci. USA 78:1023-1027.
16.   Brown, E.H., Iqbal, M.A., Stuart, S., Hatton, K.S., Valinsky, J., and
       Schildkraut, C.L. (1987) Molec. Cell Biol. 7:450-457.
17.   Henthorn, P., Mager, D., and Smithies, O. (1985) Proc. Nat'l Acad.
       Sci. USA 83:5194-5198.
18.   Grosveld, F., Taramelli, R., Kioussis, D., Vanin, E., Bartram, K., and
       Groffen, J. (1984) Nucleic Acids Res. 14:7017-7029.
19.   Ravetch, J., Siebelist, U., Korsmeyer, S., Waldmann, T., and Leder, P.
       (1981) Cell 27:583-591.
20.   Yagi, M., Gelinas R., Elder, J.T., Peretz, M., Papayonnopoulou, T.
       and Stamatoyannopoulos, G., and Groudine, M.T. (1986) Molec. Cell
       Biol. 6:1108-1116.
21.   Bentley, D. and Groudine, M. (1986) Molec. Cell Biol. 6:3481-3488.
22.   Epner, E. (1985) Ph.D thesis Columbia University.
23.   Braunstein, J., Schulze, D., Del Giudice, T., Furst, A., and
       Schildkraut, C. (1982) Nucleic Acids Res. 10:6887-6902.
24.   Weaver, D.T., Fields-Berry, S. and Depamphilis, M. (1985) Cell
       41:565-575.
25.   Germino, J. and Bastia, D. (1981) Cell 23:681-687.
26.   Edenberg, H.J. and Huberman, J.A. (1975) Ann. Rev. Genet. 9:245-
       284.
27.   Hernandez, P., Lamm, S.S., Bjerknes, C.A., and Van Hof, J. (1988)
       EMBO J. 7:303-308.
28.   Atlung, T., Rasmussen, K.V., Clausen, E.S., and Hansen, F.G. (1985)
       Role of the dnaA protein in control of replication. In M. Schaechter,
       F.C. Neidhardt, J. Ingraham, and N.O. Kjelgaard (eds): "Molecular
       Biology of bacterial growth." Boston: Jones and Bartlett. pp 282-297.
29.   Grosveld, F., van Assendelft, G.B., Greaves, D.R., and Kollias, G.
       (1987) Cell 51:975-985.
30.   Kolias, G., Wrighton, N., Hurst, J., and Grosveld, F. (1986) Cell
       46:89-94.
31.   Magram, J., Chada, K., and Constantini, F. (1985) Nature 315:338-
       340.
32.   Townes, T., Lingrel, J., Chen, H., Brinster, R., and Palmiter, R.
       (1985) EMBO J. 4:1715-1723.
33.   Wright, S. deBoer, E., Grosveld, F., and Flavell, R.A. (1983) Nature
       305:331-336.
34.   Kollias, G., Hurst, J., deboer, E., and Grosveld, F. (1987) Nucleic
       Acids Res. 15:5739-5747.

35. Wright, S., Rosenthal, A., Flavell, R.A., and Grosveld, F. (1984) Cell 38:265-273.
36. Charnay, P., Treisman, R., Mellon, P. Chao, M., Axel, R., and Maniatis, T. (1984) Cell 38:251-263.
37. Behringer, R., Hammer, R., Brinster, R., Palmiter, R., and Townes, T. (1987) Proc. Natl. Acad. Sci. USA 84:7056-7060.
38. Southern, P. and Berg, P. (1982) J. Molec. Applied Genet. 1:327-339.
39. Chao, M., Mellon, P. Charnay, P., Maniatis, T., and Axel, R. (1983) Cell 32:483-493.
40. Treisman, R., Green, M., and Maniatis, T. (1983) Proc. Nat'l Acad. Sci USA 80:7428-7432.
41. Banerji, J., Rusconi, S., and Schaffner, W. (1981) Cell 27:299-308.
42. Pellicer, A., Wagner, E., el Kareh, A., Dewey, M., Reuser, A., Silverstein, S., Axel, R., and Mintz, B. (1980) Proc. Natl Acad Sci USA 77:2098-2102.
43. Hsiung, N., Roginsky, R., Henthorn, P., Smithies, O., Kutcherlapati, R., and Skoultchi, A. (1982) Mol. Cell Biol. 2:401-411.
44. Mantei, N., Boll, W., and Weissman, C. (1979) Nature 281:40-46.
45. Stalder, J., Groudine, M., Dodgson, J., Engel, D., and Weintraub, H. (1980) Cell 19: 973-980.
46. Groudine, M., Peretz, M., and Weintraub, H. (1981) In Stamatoyannopoulos, G., Neinhuis, A (eds): "Organization and Expression of Globin Genes" New York: Alan Liss pps 163-173.
47. Weintraub, H., Beug, H., Groudine, M., and Graff, T. (1982) Cell 28:931-940.
48. Weintraub, H. (1985) Cell 42:705-711.
49. Wall, L., deBoer, E., and Grosveld, F. (1988) Genes Dev. 2:1089-1100.
50. Evans, T., Reitman, M., and Felsenfeld, G. (1988) Proc. Natl Acad Sci USA 85:5976-5980.
51. Driscoll, M.C., Dobkin, C., and Alter, B.P. (1987) Blood 70:74a.

Hemoglobin Switching, Part A: Transcriptional Regulation, pages 37–46
© 1989 Alan R. Liss, Inc.

THE β-GLOBIN DOMINANT CONTROL REGION

David R. Greaves, Mike Antoniou, Greet Blom van Assendelft, Phil Collis, Niall Dillon, Olivia Hanscombe, Jacky Hurst, Michael Lindenbaum, Dale Talbot and Frank Grosveld
Laboratory of Gene Structure and Expression
National Institute for Medical Research, London

INTRODUCTION

The strongest evidence for the existence of an important control in the flanking region of the globin gene domain was provided by the analysis of human $\gamma\beta$-thalassaemias (van der Ploeg et al., 1980; Curtin and Kan, 1986). Patients with heterozygous Dutch $\gamma\beta$-thalassaemia have a deletion that removes 100kb of DNA, leaving the β-globin gene and the promoter and enhancer regions intact. However, it abolishes expression of the deleted chromosome and leaves the gene in an inactive chromatin configuration (Kioussis et al., 1983; Wright et al., 1984; Taramelli et al., 1986). The wild-type allele on the other chromosome is expressed at normal levels, indicating that there is no shortage of trans-acting factors. This suggests a cis effect on β-globin gene transcription, which could be caused by a loss of positive acting elements or by the juxtaposition of the intact β-globin gene and sequences that remain in an inactive chromatin configuration in erythroid cells. The first indication that positive acting sites may be involved in activation of the β-globin domain came with the observation of erythroid specific DNaseI hypersensitive sites that map 6-18kb upstream from the ϵ-globin gene (Fig. 1; Tuan et al., 1985; Forrester et al., 1987 Grosveld et al., 1987).

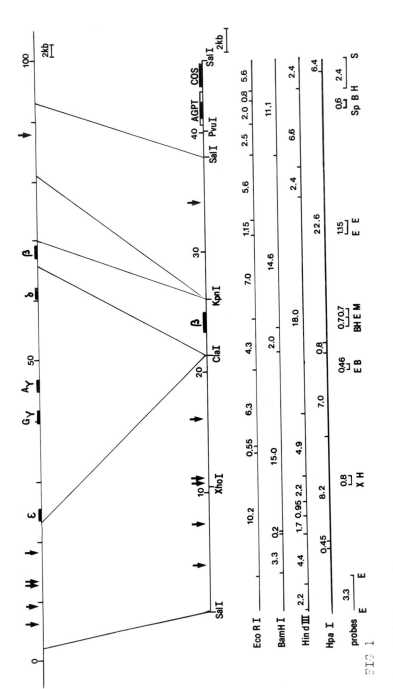

FIG 1

The human β-globin gene locus.

The human β-globin domain with all the functional genes is illustrated at the top. The β-globin minilocus leading to full expression of the β-globin gene in transgenic mice and MEL cells (Grosveld et al., 1987; Blom van Assendelft, 1988) is shown at the bottom; arrows indicate DNaseI super hypersensitive sites; sizes are in kilobases.

RESULTS

When this region is added to a human β-globin gene construct, it results in very high levels of human β-globin gene expression in transgenic mice, which is related to the copy number and independent of the integration site of the transgene (Fig. 1; Grosveld et al., 1987). The level of human β-globin gene expression observed in fetal liver RNA of 13.5 day old embryos is far higher than that observed previously in experiments with transgenic mice (Magram et al., 1985; Townes et al., 1985; Kollias et al., 1986). Furthermore, the level of human β-globin gene expression per gene copy is very nearly equal to that of the mouse endogenous β^{maj} globin gene. We are, therefore, confident that the β-globin minilocus construct contains all the positive acting sequences required for activation of the human β-globin gene during development. When the same β-globin minilocus is transfected directly into erythroid cells (MEL and K562), high level β-globin expression is obtained (Blom van Assendelft et al., 1988). The expression level per human β-globin gene is at a similar level to that of the endogenous mouse globin genes in the MEL cells. This effect is not obtained in non erythroid L-cells, where low levels of expression are obtained similar to those obtained without flanking regions. This implies that the β-globin locus does not need to undergo a complete differentiation programme to be expressed at high levels. The presence of the flanking region also results in a dramatic stimulation of transcription of the non erythroid specific promoter of the HSV thymidine kinase (tk) gene after differentiation of the MEL cells. This dramatic stimulation of the TK neo[r] gene on erythroid differentiation is not dependent on the presence of the β-globin gene promoter or enhancer sequences, as the β-globin gene can be replaced in the minilocus by a murine-human Thy-1 gene fragment and a similar stimulation of TK neo transcription is seen (Fig. 2 and see below). A similar stimulation of an SV40-neo gene is seen when it is integrated into the human β-globin locus by homologous recombination (Nandi et al., 1988). Recloning of the upstream hypersensitive sites on

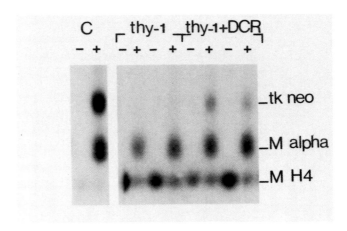

Fig. 2:

Expression analysis of the human β-globin and Thy-1 MEL clones and populations by Northern blotting.

A mouse-human hybrid Thy-1 gene was cloned between the ClaI and KpnI sites of the β-globin minilocus cosmid (lanes Thy-1 + DCR) or the cosmid vector pTCF (lanes Thy-1) and G418 resistant MEL cell populations were generated.

Fifteen micrograms of RNA from uninduced (-) as well as induced (+) clones and populations was Northern blotted after gel electrophoresis and hybridized with a mixture of the TK neo probe, the mouse α BamHI second exon probe (M alpha) and the mouse histone H4 probe (MH$_4$). RNA from untransfected MEL-C88 cells transfected with the β-globin minilocus was used as a positive control (C).

small fragments (6kb total size) shows that both the β-globin and tk-neo gene are still expressed highly (Fig. 3; Talbot *et al.*, 1988), in the case of the β-globin gene, at levels even higher than in the minilocus (Blom van Assendelft *et al.*, 1988) and this may be caused by a distance effect or by the deletion of the 3' hypersensitive region. The very high transcription levels are independent of the orientation and relative positions of the β-globin gene and the DCR and, in this respect, the DCR shows the same properties as a classical enhancer (Talbot *et al.*, 1988). Deletion of the β-globin gene leaves the tk-neo gene fully active and deletion mapping of the DCR using both MEL cells and transgenic mice shows that the most important sites in the DCR are the hypersensitive sites 2 and 3 (Collis *et al.*, unpublished). Each of these have an approximately equal contribution to the stimulation of β-globin transcription when they are tested in isolation. We have found that the β-globin gene enhancer can be deleted from these small constructs without any adverse effect on the transcription levels of the β-globin gene (Antoniou *et al.*, unpublished). We note in passing that there are no in vivo data to indicate the importance of either the γ- or β- 3' flanking enhancers. The DCR has also been shown to act on the embryonic and fetal globin genes in embryonic stem cells, MEL cells, K562 cells and transgenic mice (Lindenbaum, Dillon and Hurst, unpublished). Moreover, it can activate the α-globin gene and other non erythroid genes such as Thy-1 in MEL cells and transgenic mice (Blom van Assendelft *et al.*, 1988; Hanscombe *et al.*, unpublished). Interestingly, all of the DNaseI hypersensitive sites have already formed before the induction of MEL cell differentiation and the start of globin transcription (Blom van Assendelft *et al.*, 1988). In L-cell populations stably transfected with the β-globin minilocus construct, the only DNaseI hypersensitive site reformed is site 3. The significance of this is presently unclear, but it suggests that the formation of site 3 may be dependent upon it being present in "active" chromatin. This is assured in transfection experiments by virtue of G418 selection. Together with the data described above, this suggests that the DCR contains elements of more than one function whose action is required or affected at different times during the differentiation of erythroid cells. The DCR chromatin structure has already changed

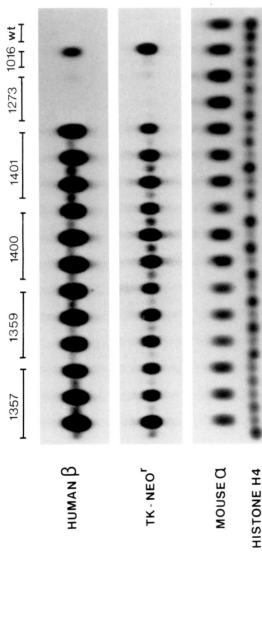

Fig. 3:

Northern blot analysis of MEL cell populations stably transfected with the human β-globin microlocus (Talbot et al., 1988).

The microlocus constructs contained the four 5' hypersensitive sites linked to the 4.8kb BglII fragment of the β-globin gene (2.0kb, 1.9kb, 1.5kb and 1kb, respectively, and the β-globin gene). 1359 contains the HSS, β-globin gene and TK-neo gene in a sense orientation from 5' to 3' in the construct. 1400 has the HSS inverted, 1401 has the β-globin gene inverted and 1357 has both the HSS and β-globin gene inverted. After electrophoresis of 10μg of RNA and blotting to nitrocellulose, the filters were probed with either a human β-globin probe or a tk-neo resistance gene probe. The filter initially probed with human β-globin was reprobed with both the mouse histone H4 probe and the mouse a-globin probe. Lanes marked 1016 contain RNA from a MEL cell clone stably transfected with the human β-globin "minilocus." Wt contains RNA from untransfected MEL C88 cells.

before the induction of differentiation, while a second inducible function is linked to the transcription of the genes.

DISCUSSION

        The control of the globin gene expression by hypersensitive region is completely dominant and the sites are present at all stages of erythroid development (Tuan et al., 1985; Grosveld et al., 1987; Forrester et al., 1987). There is a good possibility that this region controls the accessibility of the β-globin locus to trans-acting factors (see γβ-thalassaemia, Kioussis et al., 1985), perhaps like the border regions flanking the DNaseI-sensitive domain originally described for the ovalbumin genes and chicken globin genes (Lawson et al., 1982; Stadler et al., 1980). We originally speculated that these borders may contain nuclear matrix binding sites (Gasser and Laemmli, 1986) and Jarman and Higgs (1988) have recently shown the presence of two such sites immediately upstream of the 5' hypersensitive sites and downstream of the 3' hypersensitive sites. However, they found a number of other sites in the β-globin domain and it is presently not clear which of these are functionally important. The DCR might also contain one or several enhancer-like sequences, which can exert their effect over very large distances (>50kb), without stimulating the most proximal promoters in the foetal and adult stages. Interestingly, a similar control regions has recently been identified for a T-cell specific gene CD2, albeit at the 3' side of that gene (Greaves et al., 1988).

        We have now identified at least four different control regions in the β-globin gene; the DCR, the promoter and two enhancer sequences and at least two in the γ-globin gene, the DCR and the promoter. On the basis of this and the in vivo mutations, we propose the following model for the control of globin gene expression during development. The dominant control region determines the activity of the locus. In inactive non erythroid tissues there are no hypersensitive sites and the genes are not accessible to trans-acting factors. In erythroid cells the DCR becomes hypersensitive (possibly not requiring replication, Baron and Maniatis, 1986) and renders the chromatin accessible to factors. The region made accessible may be delineated by nuclear matrix binding sites. Action of the dominant control region then mediates the binding of _trans_-acting factors to regulatory promoter and enhancer sequences immediately surrounding the globin genes. The latter process largely determines the stage-specific expression and in the case of the β-globin gene, this process would involve possible negative regulatory factors and sequences to suppress the gene at early stages and positive factors acting on the enhancers and promoter to set up an extremely active transcription complex in adult stages.

However, when an isolated β-globin gene is present in the minilocus, it gives rise to a low, but still significant, level of expression in foetal cells (10% of γ-globin levels, Blom van Assendelft _et al._, 1988). Conversely, a similar situation is found for the γ-globin gene at adult stages of expression (Dillon _et al._, unpublished). When this result is compared to the transcription levels of the γ- and β-globin genes observed in a number of globin disorders, an interesting additional mechanism of fine regulation can be envisaged. In some of the non-deletion HPFH phenotypes (see above and Poncz _et al._, 1988) single point mutations in the γ-globin promoter results in elevated levels of γ-globin gene transcription and a concomitant decrease in β-globin gene transcription. Conversely, deletion of the β-globin gene promoter (Atweh _et al._, 1987; Anand _et al._, 1988) correlates with an elevated level of expression of the γ-globin gene. These _in vivo_ observations suggest that the expression of the γ- and β-globin genes influence each other and we suggest that the explanation for this phenomenon could be competition between the γ- and β-globin genes within a single domain. This competition would not be for any factor _in trans_ because competition is not observed when many copies of the β-globin gene are introduced in the same cell (Grosveld _et al._, 1987; Blom van Assendelft _et al._, 1988). The genes would therefore compete for some element _in cis_ (perhaps the DCR on the β-globin gene enhancers) when they are present in the same domain.

Such a competition has been observed in transient expression assays with the chicken adult and embryonic globin genes (Choi and Engel, 1988). If this is true, each of the genes, i.e. the γ- or β- globin genes, would have a high efficiency of expression at each of their optimal developmental stages and have a significant, but lower, efficiency (10-20%) at the inappropriate stages of development if single genes are present in the domain. However, this efficiency is decreased even further by the presence (and competition) of a more efficient gene in the same domain.

ACKNOWLEDGEMENTS

We are grateful to Cora O'Carroll for typing this manuscript.

REFERENCES

Anand R, Brehm C, Kazazian H and Vanin E (1988). Blood 172:636-641.

Atweh G, Zhu X, Buckner H, Dowling C, Kazazian H, Forget B (1987). Blood 70:1470-1474.

Baron M, Maniatis T (1986). Cell 46:591-602.

Blom van Assendelft M, Grosveld F, Greaves DR (1989). Cell. In press.

Choi O, Engel J (1988). Cell 55:17-26.

Curtin P, Kan Y (1986). In Hemoglobin Switching Fifth Conference.

Forrester W, Takegawa S, Papayannopoulou T, Stamatoyannopoulos G, Groudine M (1987). Nucl. Acids Res. 15:10159-10177.

Gasser S, Laemmli U (1986). Cell 46:521-530.

Greaves DR, Wilson FD, Lang G, Kioussis D (1988). Cell. In press.

Grosveld F, Blom van Assendelft G, Greaves D, Kollias G (1987). Cell 51:21-31.

Jarman A, Higgs D (1988). EMBO J. 7:3337-3344.

Kioussis D, Vanin E, deLange T, Flavell RA, Grosveld F (1983). Nature 306:662-666.

Kollias G, Wrighton N, Hurst J, Grosveld F (1986). Cell 46:89-94.

Lawson G, Knoll B, March C, Woo S, Tsai M, O'Malley B (1982). J. Biol. Chem. 257:1501-1507.

Magram J, Chada K, Costantini F (1985). Nature 315:338-340.

Nandi A, Roginski R, Gregg R, Smithies O, Shoultchi A (1988). Proc. Natl. Acad. Sci. USA 85:3845-3849.

Poncz M, Henthorn P, Stoeckert C, Surrey S (1988). Oxford Surveys on Eukaryotic Genes. In press.

Stadler J, Larsen A, Engel J, Dolan M, Groudine M, Weintraub H (1980). Cell 20:451-460.

Talbot D, Collis P, Antoniou M, Grosveld F, Greaves DR (1988). Submitted.

Taramelli R, Kioussis D, Vanin E, Bartram K, Groffen J, Hurst J, Grosveld F (1986). Nucl. Acids Res. 137:2098-2092.

Townes T, Lingrel J, Chen H, Brinster R, Palmiter R (1985). EMBO J. 4:1715-1723.

Tuan P, Solomon W, Qiliang L, Irving M (1985). Proc. Natl. Acad. Sci. USA 22:6384-6388.

van der Ploeg LHT, Konings M, Oort D, Roos L, Bernini L, Flavell RA (1980). Nature 283:637-642.

Wright S, Rosenthal A, Flavell RA, Grosveld FG (1984). Cell 38:265-273.

Hemoglobin Switching, Part A: Transcriptional Regulation, pages 47–61
© 1989 Alan R. Liss, Inc.

DNASE I SUPER-HYPERSENSITIVE SITES DIRECT HIGH LEVEL ERYTHROID EXPRESSION OF HUMAN α-, β- and β^s-GLOBIN GENES IN TRANSGENIC MICE

Tim M. Townes, Thomas M. Ryan, Richard R. Behringer, Richard D. Palmiter and Ralph L. Brinster

Department of Biochemistry (T.M.T., T.M.R), Schools of Medicine and Dentistry, University of Alabama at Birmingham, Birmingham, Alabama 35294. Department of Biochemistry (R.D.P.), Howard Hughes Medical Institute, University of Washington, Seattle, Washington 98195. Laboratory of Reproductive Physiology (R.R.B., R.L.B), School of Veterinary Medicine, University of Pennsylvania, Philadelphia, Pennsylvania 19104.

INTRODUCTION

The human β-like globin genes are precisely regulated in three important ways. They are expressed only in erythroid tissue, only during defined stages of development and they are expressed at high levels. To determine the sequences responsible for these three levels of control, we and others have microinjected human globin gene constructs into fertilized mouse eggs and analyzed their expression in animals that developed (Townes et al. 1985a,b; Costantini et al. 1985; Kollias et al. 1986a,b). These studies have demonstrated that sequences upstream, within and downstream of the human β-globin gene are involved in adult, erythroid-specific expression (Behringer et al. 1987; Kollias et al. 1987; Trudel and Costantini 1987).

Although transgenic mice that express human β-globin mRNA at levels equivalent to mouse β-globin mRNA have been produced with relatively small human β-globin gene constructs, many of the animals express the transgene at low levels and others do not express the gene at all. Also, the highest expressors generally have the highest number of transgenes per cell. Recently Grosveld et al. (1987) demonstrated that high levels of human β-globin gene expression can be obtained in animals with a single copy of the transgene if sequences at the extreme ends of the human β-globin locus are included in the injected constructs. Several years ago, Tuan et al. (1985) and Forrester et al. (1986) mapped sites that were super-hypersensitive to DNase I digestion in these regions. These DNase I super-hypersensitive sites (HS) are located 6-22 kb upstream of the ε-globin gene and 19 kb downstream of the β-globin gene. The sites are present specifically in erythroid tissue at all stages of development. Fig. 1 depicts the location of these sites in the human β-globin locus.

The structure of mutant loci from patients with several hemoglobinopathies suggests that the upstream super-hypersensitive sites are required for efficient β-globin gene expression in humans (Bunn and Forget 1986, Stamatoyannopoulos et al. 1987). Hispanic (Driscol, pers comm), English and Dutch γδβ thalassemias result from deletions that remove all or most of the upstream HS sites (Fig. 1). Although the β-globin gene is intact in these patients, no β-globin mRNA is produced from the mutant alleles. On the other hand, several other deletions suggest that the downstream site may not be essential. High levels of human γ-globin gene expression are observed in patients with two deletion forms of hereditary persistence of fetal hemoglobin (HPFH-1 and HPFH-2). Both of these mutant alleles lack the downstream super-hypersensitive site (HSVI; see Fig. 1). To determine whether the downstream site is required for high level expression of the human β-globin gene, we have made constructs containing all of the upstream sites but lacking the downstream site. These constructs were microinjected into fertilized mouse eggs and the levels of human β-globin mRNA

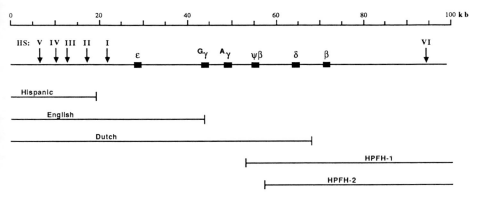

Fig. 1 Human β-globin locus illustrating DNase I super-hypersensitive sites and various HPFH and γδβ thalassemic deletions.

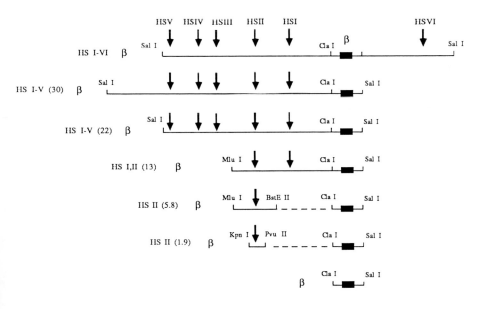

Fig. 2 HS β and β constructs injected into fertilized mouse eggs. Numbers in parentheses represent the length of flanking sequence in kilobase pairs. The β-globin gene in all of these constructs is on a 4.1 kb fragment containing 815 bp of 5' and 1700 bp of 3' flanking sequence.

were analyzed in mice that developed. Constructs
containing only HS I and II or HS II alone were
also tested to determine the minimum number of
super-hypersensitive sites required for high level
expression of the human β-globin gene in transgenic
mice.

HS β-globin transgenic mice.

    Figure 2 illustrates the 7 constructs that were
purified from vector sequences and injected into
fertilized mouse eggs (Brinster et al., 1985).
These eggs were transferred into the uteri of
psuedo-pregnant foster mothers and after 16 days of
development the embryos were removed. Total
nucleic acids were prepared from the erythroid
fetal livers and from brains, and transgenic mice
were identified by DNA dot hybridization with β-
globin and HS II specific probes. Fetal liver DNA
from positive animals was then analyzed by Southern
blotting to determine transgene copy number and
integrity. Fifty-one transgenic animals that
contained intact copies of the transgene were
analyzed for expression by solution hybridization
with oligonucleotide probes as described previously
(Townes et al. 1985). A summary of these results
is presented in Table I. In this study only 7 of
23 animals without HS sites expressed the
transgene. In contrast, 50 of 51 animals that
contained HS sites inserted upstream of the human
β-globin gene expressed correctly initiated human
β-globin mRNA in fetal liver and no expression was
detected in fetal brain. These results, like the
results of Grosveld et al. (1987) with a construct
containing HS I-VI β, suggest that the HS sites
activate expression regardless of the site of
transgene integration. However, expression is not
totally position independent. The range of
expression varied widely with all of the constructs
tested and levels of human β-globin mRNA were not
absolutely correlated with transgene copy number.
Nevertheless, the average levels of expression per
gene copy were high for all of the HS β-globin
constructs tested. The HS I-V (30) β and HS I-V
(22) β constructs were expressed at an average

Table I. Summary of HS β transgene expression

Transgene	Fraction expressors	Percent endogenous m β-globin mRNA [*]	Percent expression per gene copy [+]	
			Mean	(Range)
HS I-VI β	3/3	5 - 26	52	(20 - 84)
HS I-V (30) β	13/13	18 - 316	108	(16 - 200)
HS I-V (22) β	9/9	1 - 380	109	(2 - 208)
HS I,II (13) β	13/13	9 - 347	49	(9 - 92)
HS II (5.8) β	6/7	8 - 108	40	(6 - 84)
HS II (1.9) β	4/4	56 - 194	40	(13 - 63)
β	7/23	0.2 - 23	0.3	(0.1 - 0.6)

[*] $\left(\dfrac{h\,\beta\ mRNA}{m\,\beta\ mRNA} \times 100 \right)$ [+] $\left(\dfrac{h\,\beta\ mRNA\,/\,h\beta\ gene}{m\beta\ mRNA\,/\,m\beta\ gene} \times 100 \right)$

Human and mouse β-globin mRNA levels were quantitated by solution hybridization with human β- and mouse β-globin specific oligonucleotides. The values of percent expression per gene copy were calculated assuming 4 mouse β-globin genes/cell. The (C57BL/6 x SJL) F2 mice used in this study have the Hbb[s] or single haplotype. The β-globin locus in this haplotype contains 2 adult β-globin genes (β[s] and β[t]) per haploid genome. These mice also have 2 α-globin genes (α1 and α2) per haploid genome. Copies/cell of HS β transgenes were determined by densitometric scanning of the Southern blots (data not shown).

level of 108% and 109%, respectively, of endogenous mouse β-globin per gene copy and all other HS β constructs were expressed at 40-49% of endogenous mouse β-globin per gene copy. This high level of expression was obtained even when a 1.9 kb fragment containing only HS II was inserted upstream of the human β-globin gene. The average level of expression per gene copy for a human β-globin construct that did not contain HS sites was only 0.3% of endogenous mouse β-globin. This average

level of expression is 133-363 times lower than
constructs containing HS sites. Finally, we
suspect that the average level of expression for
the HS I-VI β construct was lower than 100% per
gene copy because only 3 animals were obtained.

HS α-globin transgenic mice

We have attempted for some time to obtain
transgenic mice that express the human α-globin
gene. We tested the human α1-globin gene on a 3.8
kb BglII-EcoRI fragment, the α2 and α1 genes on a
14 kb BamHI fragment, the entire human α-globin
locus on a 42 kb cosmid fragment and the human α-
and β-globin genes in various orientations on the
same fragment. Although 61 animals that contain
intact copies of these transgenes were obtained, no
α-globin mRNA was detected in any tissue.
Since α-globin genes with small or large
amounts of flanking sequence were not expressed in
transgenic mice, we inserted HS I and II from the
human β-globin locus immediately upstream of the
human α1-globin gene. This HS I,II α-globin
construct was injected into fertilized eggs and
expression was analyzed in 16 day fetal liver and
brain of animals that developed. Figure 3
illustrates a primer extension analysis of fetal
liver RNA from 4 animals that contained the α-
globin gene alone and 12 animals that contained HS
I,II α-globin. No human α-globin mRNA could be
detected in the 4 transgenic mice containing the α-
globin gene alone or in the one animal (5058) that
had lost the HS II site (data not shown). However,
all 11 transgenic mice that contained intact copies
of HS I,II α-globin expressed correctly initiated
human α-globin mRNA. Accurate quantitative values
of human α- and mouse β-globin mRNA levels were
determined by solution hybridization with human α-
and mouse β-specific oligonucleotides. No human α-
globin mRNA could be detected in α transgenic
animals or in the one HS II-minus mouse at a
sensitivity of 0.2% of endogenous mouse β-globin
mRNA. However, mice that contained intact copies
of HS I,II α expressed human α-globin mRNA at
levels ranging from 4% to 337% of mouse β. When
human α- and mouse β-globin mRNA levels were

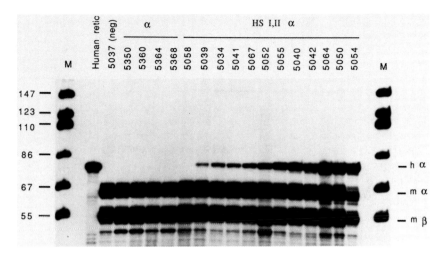

Fig. 3 Primer extension analysis of fetal liver RNA from α-globin and HS I, II α-globin transgenic mice. The authentic human α-globin primer extension product (hα) is 76 bp and the correct mouse α- and β-globin products (mα, mβ) are 65 and 53 bp, respectively.

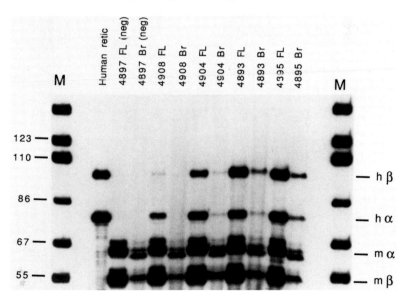

Fig. 4 Primer extension analysis of fetal liver and brain RNA from HS I,II α-globin/ HS I,II β-globin transgenic mice. The authentic human β-globin primer extension product (hβ) is 98 bp.

calculated per gene copy, human α values ranged
from 28.8% to 90.0% of endogenous mouse β.
Expression was originally calculated as a
percentage of mouse β instead of mouse α-globin mRNA
in case the endogenous mouse α-globin gene was down
regulated. However, the primer extensions in
figure 3 and subsequent solution hybridizations
with a mouse α-globin specific oligonucleotide
demonstrate that mouse α-globin is not consistently
down regulated even in high expressors and that
mouse α- and β-globin mRNA levels are essentially
equivalent.

Co-expression of Human α- and β-globin genes

        The results described above demonstrate that
both the human α- and β-globin genes are expressed
at high levels in erythroid tissue of transgenic
mice when these genes are flanked by hypersensitive
sites from the human β-globin locus. These results
suggested that a complete human hemoglobin could be
synthesized in transgenic mice if fertilized eggs
were coinjected with human α- and β-globin genes
flanked by super-hypersensitive sites.   Therefore,
we injected equimolar amounts of the HS I,II α-
globin and HS I,II β-globin constructs into
fertilized mouse eggs. These eggs were transferred
into the uteri of psuedo-pregnant foster mothers
and after 16 days development the fetuses were
removed.   Fetal liver and brain RNA of the 4
animals that contained intact copies of the
transgenes were then analyzed for correctly
initiated human α- and β-globin mRNA by primer
extension. The results are illustrated in Fig. 4.
Lane 1 is human reticulocyte RNA and lanes 2 and 3
are fetal liver (FL) and brain (Br) RNA from a non-
transgenic mouse control. Fetal liver and brain
RNA from the 4 transgenic animals are analyzed in
lanes 4-11. All 4 of the transgenic animals
expressed correctly initiated human α- and human β-
globin mRNA specifically in fetal liver. The small
amount of human α- and β-globin mRNA in the brain
results from blood contamination because equivalent
amounts of mouse α- and mouse β-globin mRNA are
also observed in this non-erythroid tissue.

Table II. Quantitation of HS I,II α and HS I,II β coexpression

Mouse	4908	4904	4893	4895	5393
h β gene (copies/cell)	3.0	2.0	3.0	6.0	5.0
h α gene (copies/cell)	3.0	2.0	1.0	3.0	4.0
$\dfrac{\text{h}\beta\ \text{mRNA}}{\text{m}\beta\ \text{mRNA}} \times 100$	0.7	12	51	49	69
$\dfrac{\text{h}\alpha\ \text{mRNA}}{\text{m}\beta\ \text{mRNA}} \times 100$	7.7	31	30	52	61
$\dfrac{\text{h}\beta\ \text{mRNA}\ /\ \text{h}\beta\ \text{gene}}{\text{m}\beta\ \text{mRNA}\ /\ \text{m}\beta\ \text{gene}} \times 100$	1.0	24	68	33	55
$\dfrac{\text{h}\alpha\ \text{mRNA}\ /\ \text{h}\alpha\ \text{gene}}{\text{m}\beta\ \text{mRNA}\ /\ \text{m}\beta\ \text{gene}} \times 100$	10	62	120	69	61

Human and mouse α- and β-globin mRNA levels were quantitated by solution hybridization with oligonucleotides specific for human α- and β-globin and mouse α- and β-globin specific oligonucleotides. The values of percent expression per gene copy were calculated as described in the legend to Table I.

Accurate quantitative values of human α-, human β- and mouse β-globin mRNA levels were determined by solution hybridization with human α- , human β- and mouse β-specific oligonucleotides. The data in Table II demonstrate that human α- and β-globin mRNA levels as high as 120% and 68% per gene copy, respectively, of endogenous mouse β-globin mRNA were obtained. These results demonstrate that high levels of human α- and β-globin gene expression can be achieved in transgenic mice after coinjection of HS I,II α- and β-globin constructs.
        Transgenic mice that express high levels of human α- or β-globin mRNA may not survive to birth. Death in utero would presumably result from the deleterious effects of excess α- or β-globin polypeptides in erythroid cells. High levels of excess α-globin polypeptides in β-thalassemic patients form aggregates that precipitate inside the cell. These precipitates disrupt the

erythrocyte membrane and lead to premature red cell death. Precipitates of β_4 tetramers in α-thalassemia lead to similar problems (Bunn and Forget 1986). The high level coexpression of human α- and β-globin genes in the animals described above suggested that live born animals could be obtained after coinjection of HS I,II α- and β-globin constructs. Therefore, we injected equimolar amounts of HS I,II α-globin and HS I,II β-globin gene constructs into fertilized mouse eggs and analyzed the blood of adult transgenic animals that developed. A non-denaturing isoelectric focusing (IEF) gel of hemolysates from 2 HS I,II α- and β-globin mice is illustrated in Fig. 5A. The first lane is a non-transgenic mouse control and the last lane is a normal human sample. The predominant band in each of the controls is the major adult hemoglobin; mouse $\alpha_2\beta_2$ or human $\alpha_2\beta_2$, respectively. In both transgenic mouse samples 5394 and 5393, bands that run at the same pI as human HbA ($h\alpha_2\beta_2$) and mouse HbA ($m\alpha_2m\beta_2$) are observed. In mouse 5393 human HbA appears to be the predominant hemoglobin. In addition to human and mouse hemoglobins, 2 other major bands are observed in both samples.

To determine the composition of these novel bands and to confirm the identity of human and mouse hemoglobins, the 4 bands in sample 5393 were excised from the gel. Protein was then eluted, dialyzed and analyzed by electrophoresis on a denaturing cellulose acetate strip. Figure 5B illustrates the result of this experiment. Lysates of mouse, human and 5393 blood samples were run as controls. Mouse α-globin and β-globin polypeptides (lane 1) as well as human α- and β-globin polypeptides (lane 2) are well separated on the cellulose acetate strip. Lane 3 demonstrates that all 4 polypeptides are present in sample 5393. A densitometric scan of this lane demonstrates that the levels of human α- and β-globin polypeptides are 105% and 110%, respectively, of mouse α- and β-globin. Lane 1 illustrates the banding pattern observed when the top band of sample 5393 in Fig. 5A was run on the denaturing strip. This band is composed of human α- and mouse β-globin chains.

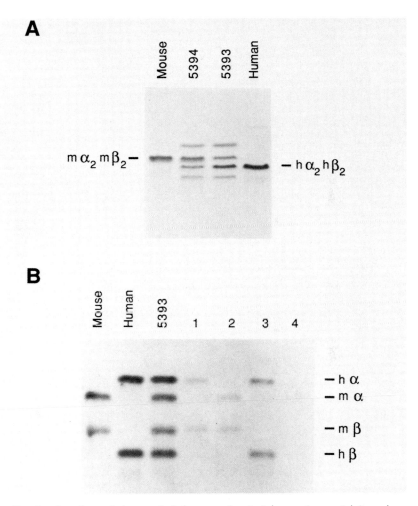

Fig. 5 Analysis of hemoglobins and globin polypeptides in HS I,II α-globin/ HS I,II β-globin transgenic mice. Panel A. Hemolysates of normal mouse and human blood and blood from mice 5394 and 5393 were run on a native agarose isoelectric focusing gel and photographed without staining. Panel B. Hemoglobins were denatured in urea-mercaptoethanol, electrophoresed on cellulose acetate strips and stained with imido black. Lanes marked mouse, human, and 5393 are hemolysates of normal mouse and human blood and blood from mouse 5393. Lanes 1-4 are hemoglobins purified from individual bands of sample 5393 (top to bottom, respectively) on the isoelectric focusing gel in panel A.

The second band in Fig. 5A is mouse α- and mouse β-globin and the third band is human α- and human β-globin as expected. The polypeptides composing band 4 in Fig. 5A are barely visible in Fig. 5B but are mouse α- and human β-globin. These results clearly demonstrate that high levels of complete human hemoglobin can be produced in transgenic mice after coinjection of human HS α- and β-globin genes.

Synthesis of HbS in Transgenic Mice

The results described above suggested that high levels of human sickle hemoglobin (HbS) could be produced in transgenic mice if fertilized eggs were coinjected with HS α-globin and HS $β^S$-globin genes. Therefore, we coinjected HS I-V α-globin and HS I-V $β^S$-globin constructs into fertilized eggs and analyzed hemolysates of transgenic animals that developed. Figure 6 illustrates an IEF gel of the 3 transgenic animals that were obtained. Lanes 1 and 2 are controls. Lane 1 is a hemolysate from a normal human subject (AA) and lane 2 is a hemolysate from a patient with sickle trait (AS). Lanes 3-5 are HS I-V α /HS I-V $β^S$ transgenic mouse samples. Lane 6 is sample 5393 described above and lane 7 is a normal mouse hemolysate run as a control. All 3 transgenic mouse samples contained a band that comigrated with HbS. Although the data suggest that high levels of HbS are produced in these mice, the results are somewhat obscured because HbS and the heterotetramer h$α_2$m$β_2$ comigrate on these gels. However, primer extension analysis (data not shown) demonstrates that all 3 transgenic mice make significant amounts of both human α- and human $β^S$-globin mRNA. The levels of human α- and human $β^S$-globin mRNA in mouse 5659 are 110% and 324% of endogenouse mouse α- and β-globin mRNA. Unfortunately, preliminary experiments suggest that erythrocytes from the 5659 mice do not sickle under low oxygen tensions. Therefore, we have mated the HS I-V α- and $β^S$-globin transgenes from the 5659 line into mouse α-thalassemic (Martinell et al., 1981; Whitney et al., 1981) and β-thalassemic (Skow et al., 1983) lines and we are now mating these

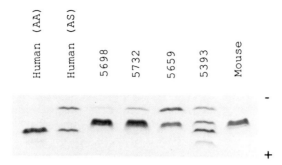

Fig. 6 Isoelectric focusing gel of hemolysates from HS I,II
α-globin/ HS I,II β^S-globin transgenic mice. Lanes 1 and 2
(AA, AS) are hemolysates from a normal human control and a
patient with sickle trait, respectively. Lanes 3-5 (5698,
5732 and 5659) are HS I,II α-globin/ HS I,II β^S-globin
transgenic mouse samples. Lanes 6 is sample 5393 from Fig.
5 and lane 7 is a (C57BL/6 x SJL) F2 mouse control.

lines to obtain animals that are homozygous for the
transgenes, homozygous for the β-thal allele and
heterozygous for the α-thal allele. Blood from
these animals will then be deoxygenated and
examined for sickled erythrocytes and HbS
polymerization will be studied in hemolysates and
in intact cells. Hopefully, the low levels of
mouse α and β^{min} in these animals will not inhibit
sickling. However, if erythrocytes do not sickle,
it will be necessary to make new α- and β-
thalassemic mouse lines from embryonic stem (ES)
cells (Mansour et al. 1988) in which both α-globin
genes and β^{maj} and β^{min} globin genes are deleted.
The HS α- and β^S transgenes can then be mated into
these thalassemic backgrounds so that the best
possible conditions for in vivo sickling can be
obtained.

REFERENCES

Behringer RR, Hammer RE, Brinster RL, Palmiter RD
and Townes TM (1987). Two 3' sequences direct adult
erythroid-specific expression of human β-globin
genes in transgenic mice. Proc Natl Acad Sci 84:
7056-7060.

Brinster RL, Chen HY, Trumbauer ME, Yagle MK and
Palmiter RD (1985). Factors affecting the
efficiency of introducing foreign DNA into mice by
microinjecting eggs. Proc Natl Acad Sci 82: 4438-
4442.

Bunn HF and Forget BG (1986). Hemoglobin:
Molecular, Genetic and Clinical Aspects. W.B.
Saunders Co. Philadelphia.

Costantini F, Radice G, Magram J,
Stamatoyannopoulos G and Papayannopoulou T (1985).
Developmental regulation of human globin genes in
transgenic mice. Cold Spring Harbor Symp Quan Biol
50: 361-370.

Doetschman T, Maeda N and Smithies O (1988)
Targeted mutation of the Hprt gene in mouse
embryonic stem cells. Proc Natl Acad Sci USA
85:8583-8587.

Forrester W, Thompson C, Elder JT and Groudine M
(1986). A developmentally stable chromatin
structure in the human β-globin gene cluster. Proc
Natl Acad Sci 83: 1359-1363.

Grosveld F, van Assendelft GB, Greaves DR and
Kollias G (1987). Position-independent, high-level
expression of the human β-globin gene in transgenic
mice. Cell 51: 975-985.

Kollias G, Hurst J, deBoer E, Grosveld F. (1987).
The human β-globin gene contains a downstream
developmental specific enhancer. Nucleic Acids Res
15: 5739-5747.

Kollias G, Wrighton N, Hurst J and Grosveld F
(1986). Regulated expression of human $^A\gamma-$, β-, and
hybrid γ/β-globin genes in transgenic mice:
Manipulation of the developmental expression
patterns. Cell 46: 89-94.

Mansour SL, Thomas KR and Capecchi MR (1988).
Disruption of the proto-oncogene int-2 in mouse

embryo-derived stem cells: a general strategy for targeting mutations to non-selectable genes. Nature 336:348-352.

Martinell J, Whitney III JB, Popp RA, Russell LB and Anderson WF (1981). Three mouse models of human thalassemia. Proc Natl Acad Sci USA 78: 5056-5057.

Skow LC, Burkhart BA, Johnson FM, Popp RA, Popp DM, Goldberg SZ, Anderson WF, Barnett LB and Lewis SE (1983). A mouse model for β-thalassemia. Cell 40: 1043-1044.

Stamatoyannopoulos G, Nienhuis AW, Leder P and Majerus PW (1987). Molecular Basis of Blood Diseases. W.B. Saunders Company.

Thomas KR and Capecchi MR (1987). Site-directed mutagenesis by gene targeting in mouse embryo-derived stem cells. Cell 51: 503-512.

Townes TM, Chen HY, Lingrel JB, Palmiter RD and Brinster RL (1985). Expression of human β-globin genes in transgenic mice: Effects of a flanking metallothionein-human growth hormone fusion gene. Mol Cell Biol 5: 1977-1983.

Townes TM, Lingrel JB, Chen HY, Brinster RL and Palmiter RD (1985). Erythroid-specific expression of human β-globin genes in transgenic mice. EMBO J 4: 1715-1723.

Trudel M and Costantini F (1987). A 3' enhancer contributes to the stage-specific expression of the human β-globin gene. Genes and Development 1: 945-961.

Tuan D, Soloman W, Li Q and London I (1985). The β-like-globin gene domain in human erythroid cells. Proc Natl Acad Sci 82: 6384-6388.

Whitney JB III, Cobb RR, Popp RA and O'Rourke TW (1981). Deletions in the α-globin gene complex in α-thalassemic mice. Proc Natl Acad Sci 78: 7644-7647.

Hemoglobin Switching, Part A: Transcriptional Regulation, pages 63–72
© 1989 Alan R. Liss, Inc.

Characterization of a human globin enhancer element

Dorothy Y.H. Tuan, William B. Solomon, Resy
Cavallesco, George Huang, and Irving M. London
Harvard-MIT Division of Health Sciences and
Tech., and Department of Biology, Massachusetts
Institute of Technology, Cambridge, MA   02139

Introduction
        DNA mediated gene transfer experiments have shown that the human β-globin gene with its immediate 5' and 3' flanking sequences can be expressed in a tissue and developmental-stage specific manner (Magram et al, 1985; Townes et al 1985; Chada et al 1986; Kollias et al 1986; Dzierzak et al, 1988). The transcriptional efficiency of such introduced β-globin genes is, however, generally low. Efficient transcription may require cis control by sequence elements that reside outside of the globin structural genes and their immediate flanking sequences. In erythroid cells, four major DNase 1 hypersensitive sites have been mapped in the far upstream area of the β like globin gene domain. HSI at -6.1 to -6.4 Kb, HSII at -10 to -11 Kb, HSIII at -14.5 to -14.7 Kb and HSIV at -17.5 to -18 Kb 5' of the ε-globin gene (Tuan et al 1985; Forrester et al, 1987). The erythroid specificity and developmental stability of these sites suggest that they may be involved in regulation of not only the more proximal embryonic ε globin gene, but also the far distal fetal γ and adult β globin genes. The preferential accessibility of these sites to exogenous DNase 1 with respect to the rest of the DNA in the globin gene domain indicates that the DNA sequences in these sites exist in a relatively open chromosome structure and may be preferentially accessible also to trans-activating regulatory factors in erythroid cells. In response to the regulatory signals they may exhibit enhancer-like functions in stimulating the transcriptional efficiency of the otherwise poorly transcribed globin genes. We have previously shown by transient CAT assays that in 20 Kb of DNA contiguous to the 5' end of the ε-globin gene (see Fig. 1A), HSII at -10 to -11 Kb spanned by a 1.9 Kb Hind III DNA fragment, possesses transcriptional enhancer function (Tuan et al, 1987). Here we present data which show that the enhancer element resides in a 0.8 Kb fragment directly underlying

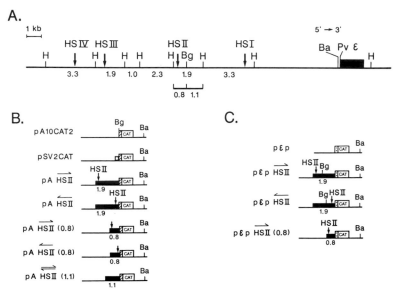

Fig. 1: Recombinant CAT plasmids containing the enhancer element in combination with either the SV40 or the ε globin promoter. A. The partial restriction map of DNA spanning four erythroid specific major DNase 1 hypersensitive sites 5' of the ε globin gene. Restriction sites used in plasmid construction are shown. H, Hind III, pv: PvuII, Ba: BamH1, Bg:BgIII. Numbers below the map denote sizes in Kb of the various Hind III fragments or subfragments. Heavy vertical arrows denote locations of the hypersensitive sites (HSI-HSIV). The filled box represents the ε-globin gene. B. Recombinant CAT plasmids containing the SV40 promoter: the hatched Box 5' of the CAT gene represents the SV40 promoter. The horizontal filled bars 5' of the SV40 promoter represent the DNA inserts to be tested for enhancer function. Horizontal arrows denote genomic (—→) or reverse genomic (←—) orientation of the inserts. pA10CAT2: the enhancerless reference plasmid. pSV2CAT: CAT plasmids containing the SV40 enhancer (represented by the unfilled box). pAHSII or pAHSII: the CAT plasmids containing the 1.9 Kb DNA fragment spanning HSII. pA HSII (0.8) or pA HSII (0.8): the 0.8 Kb DNA subfragment directly underlying HSII inserted in pA10CAT2. pA HSII (1.1): the 1.1 Kb subfragment which does not span any hypersensitive sites, inserted in either genomic (—→) or reverse genomic (←—) orientation in pA10CAT2. C. Recombinant CAT plasmids containing the ε-globin promoter. pεp: the enhancerless reference CAT plasmid containing the ε-globin promoter which is represented by the dotted box. pεp HSII or pεp HSII: the recombinant plasmid containing the 1.9 Kb enhancer fragment spliced 5' of the ε-globin promoter in pεp. pεp HSII (0.8): the 0.8 Kb enhancer fragment spliced in pεp.

HSII, and that it functions in an erythroid specific manner, and can stimulate the transcription of a cis-linked gene at the proper CAP sites. We shall also present data which suggest that the enhancer element is functional in erythroid cells at both embryonic and adult developmental stages. These findings are consistent with the proposed role of the enhancer element in transcriptional activation of the β-like globin genes throughout erythroid development.

Enhancer function is located in a 0.8 Kb DNA fragment directly underlying HSII at 10.2-11.0 Kb 5' of the ε globin gene:

In the 1.9 Kb DNA fragment exhibiting enhancer activity, HSII is located at the 5' end of the fragment (Fig. 1A). One might argue that enhancer function could reside in the 3' half of the fragment which does not span any DNase 1 hypersensitive sites. By subdividing the 1.9 Kb fragment, we show here that the enhancer function does reside in the 5'

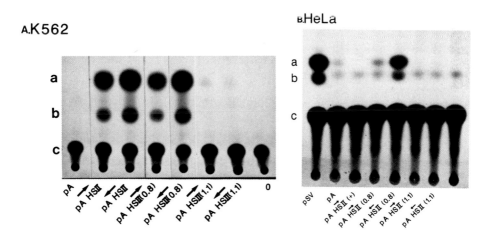

Fig. 2: Thin layer chromatography (TLC) of CAT enzymatic activity, in K562(A) and HeLa(B) cells, of recombinant plasmids containing the SV40 promoter. Each lane represents the CAT activity of the respective plasmid as marked. pA: pA10CAT2. pA HSII (+) contains the 1.9 Kb enhancer fragment spliced in genomic orientation into the Bam H1 site 3' of the CAT gene in pA10CAT2. 0: mock transfection. Acetylated chloramphenicol spots in the TLC are marked by a and b and unacetylated chloramphenicol spots by c in the left margin of the TLC.

0.8 Kb subfragment which directly underlies HSII. The recombinant CAT constructs containing the enhancer fragments linked in cis to the CAT gene and the SV40 promoter are shown in Fig. 1B. Transfected into K562 cells, which possess properties characteristic of human ery-throid cells, recombinant CAT plasmids containing the 1.9 Kb fragment spanning HSII spliced either genomic or reverse genomic orientation 5' of the CAT gene (pA HSII and pA HSII), significantly activate the cis linked CAT gene, (Fig. 2A and Table 1). Recombinant CAT plasmids containing the 0.8 Kb 5' subfragment directly underlying HSII, pA HSII (0.8) and pA HSII (0.8), similarly enhance the CAT gene. On the other hand, the CAT plasmids containing the 1.1 Kb 3' subfragment which does not span HSII, pA HSII (1.1) or pA HSII (1.1), as well as the enhancerless reference plasmid pA10CAT2, do not noticeably enhance the transcription of the CAT gene.

The above functional assays thus define enhancer activity to within a 0.8 Kb DNA fragment directly underlying HSII between a Bgl II site at -10.2 Kb and a Hind III site at -11.0 Kb 5' of the CAP site of the ε-globin gene (Fig. 1A).

Erythroid specificity of the enhancer element

To determine the erythroid specificity of the enhancer element, the above recombinant CAT plasmids were also transfected into the non-erythroid HeLa cells. In HeLa cells, the 0.8 Kb fragment underlying HSII in genomic orientation in pAHII (0.8) (Fig. 2B., Table 1), as well as the 1.9 Kb fragment in genomic orientation in pA HSII does not show enhancer activity in the non-erythroid HeLa cells. The same DNA fragments in reverse genomic orientation in pA HSII (0.8) and pA HSII (Fig. 2B, Table 1), however, stimulate CAT activity by approximately 10 fold over the enhancerless pA10CAT2. In another non-erythroid, monocytic leukemia cell line, the THP-1 cells, the 1.9 Kb DNA fragment in reverse genomic orientation in pA HSII, however, does not show enhancer function (Table 1). The above recombinant plasmids which show some enhancer activity in HeLa cells contain the CAT gene driven by the SV40 promoter. Globin promoters have been reported to confer erythroid specific expression on a linked gene (Rutherford and Nienhuis, 1987). To determine if the erythroid specificity of the enhancer can be improved, we have constructed another series of recombinant CAT plasmids, in which the enhancer element is coupled to the homologous ε-globin promoter (see Fig. 1C). Transfected into K562 cells, the CAT plasmids containing the 1.9 Kb in either genomic or reverse genomic orientation coupled to the ε-globin promoter, pεp HSII and pεp HSII, activate the CAT gene by up to 300 fold more than does the enhancerless pεp which contains only the ε globin promoter 5' of the CAT gene (Fig. 3). When transfected into non-erythroid HeLa and THP-1 cells, these same plasmids do not activate the CAT gene (Fig. 3, Table 1). Thus, in combination with the homologous globin promoter, the activity of the enhancer element and its erythroid specificity appear to be increased.

HeLa

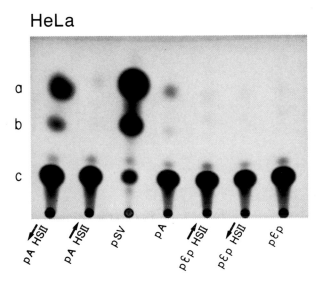

K562

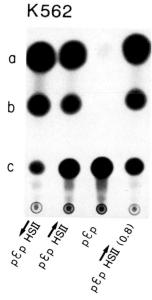

Fig. 3:   CAT enzymatic activities in HeLa and K562 cells of recombinant plasmids containing the embryonic ε-globin promoter.

Table 1

	K562	HeLa	THP-1	MEL
pA10CAT2	1	1	1	1
pSV2CAT	8.4±5.4	160±40	2.3±1.1	37±9
pA HSI1	67 ±22	8.8±5	0.4±0.25	14±7
pA HSI1	26 ±12	0.7±0.3		16±11
pA HSI1-(0.8)	88 ±45	13±5		18±8
pA HSI1-(0.8)	32 ±10	3±1.5		13±5
pA HSI1-(1.1)	1.1±0.2	0.7		
pA HSI1-(1.1)	1.7±0.3	0.7		
0	0.4			0.1

			c	
pεp	1	1	1	
pεp HSI1	270±70	1±0.1	2	
pεp HSI1	245±75	0.9±0.2	0.3	
pεp HSI1-(0.8)	248±80		0.25±.02	
0	0.7			

Table 1. Relative CAT activities of CAT plasmids driven either by the SV40 or the ε-globin promoter. Relative CAT activity of a test plasmid is defined as the percentage of total input ^{14}C-counts in ^{14}C-chloramphenicol that is converted to the acetylated forms (spots a and b on the TLC), over the percentage of conversion by the enhancerless pA10CAT2 or pεp plasmid whose CAT activity is set at 1. Averaged values of relative CAT activities presented were determined from at least two separate experiments.

To minimize variation in relative CAT values due to the variability in transfection efficiencies between experiments, we have determined the CAT enzymatic activities from CAT extracts containing approximately equal numbers of the various transfecting plasmids. We have also normalized the relative CAT enzymatic activities with reference to amounts of the expression product, the Human Growth Hormone (HGH), of a co-transfecting reference plasmid. The values of CAT activities determined by the two methods agree with each other within the experimental variations of the assay and produce consistent trends in relative CAT activities (Tuan et al, submitted).

The enhancer element activates transcription of the cis-linked CAT gene at the CAP site:

To determine the initiation sites of the CAT messages in K562 cells we carried out RNA protection assays (Melton et al, 1984) of the K562 RNA by using a single stranded RNA probe which is complementary to the CAT message and spans 220 nucleotides (nt) of the CAT coding sequence and 139 nt of upsteam sequence including the SV40 promoter. When annealed to the probe, the CAT messenger RNA's in K562 cells produced protected fragments of 321 and 326 nt (Fig.4A) and were thus initiated from the proper CAP sites of the SV40 promoter (Fig. 4B). Longer protected fragments of 342 and 347 nt which were initiated from within the TATA boxes of the promoter were also present. Furthermore, a minor band representing full-length protected probe of 359 nt was present. This observation suggests that some of the CAT transcripts were initiated upstream of the CAP sites, not only from within the TATA boxes but probably also from within the erythroid enhancer sequence (Fig. 4B). The relative amounts of the CAT messages (Fig. 4A) appear to be proportional to the CAT enzymatic activities of the respective CAT plasmids as determined by the CAT assays (Table 1).

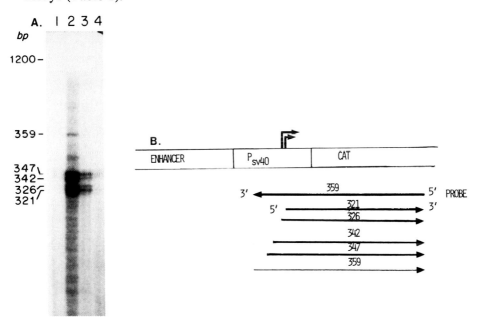

Fig. 4: A. RNA protection assays of K562 RNA: RNA from K562 cells transfected by mock transfection (lane 1); by pA HSII (0.8) (lane 2); by pA HSII (0.8) (lane 3) and by pA (lane 4).

  B. Graphic representation of the location of the probes and protected fragments in relation to the location of the SV40 promoter and the enhancer sequences. Thickness of the horizontal lines denotes relative amounts of the protected CAT transcripts.

Developmental-stage independence of the enhancer element:
    To determine if the enhancer element functions in a
developmental-stage independent manner, i.e., if it activates CAT
transcription not only in erythroid cells expressing the embryonic ε and
fetal γ globin genes as in K562 cells but also in erythroid cells at the adult
stage of development expressing the adult β-globin gene, we have
transfected various recombinant CAT plasmids into mouse
erythroleukemia (MEL) cells. MEL cells express the adult mouse β-
globin gene and possess properties characteristic of adult erythroid cells.
In MEL cells, the 1.9 Kb and the 0.8 Kb fragments coupled to the SV40
promoter in either reverse genomic or genomic orientation in pA HSII,
pA HSII, pAHSII (0.8) or pA HSII (0.8) respectively, all activate CAT
gene expression by 10 to 20 fold; the SV40 enhancer element in
pSV2CAT, on the other hand, activates the CAT gene by approximately
40 fold (Fig. 5A and Table 1). Coupled to the adult β-globin promoter,
the enhancer element activates the cis-linked CAT gene by up to 100 fold
in MEL cells (Fig. 5B Huang, G. and Tuan, D., unpublished results).

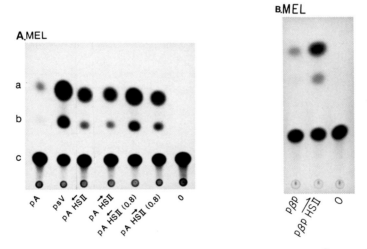

Fig. 5:   CAT enzymatic activities in MEL cells, of recombinant
               CAT plasmids containing the SV40 promoter and the
               adult β-globin promoter. pβp is the enhancerless
               PβGLCAT containing the β globin promoter (Fordis, M.
               et al 1985). pβp H̄S̄II contains the 1.9 Kb enhancer
               fragment spliced in genomic orientation 5' of the β globin
               promoter and the CAT gene in pβp.

    The above experiments suggest that the human erythroid enhancer
element is active not only in K562 cells which express the embryonic

globin genes, but also in MEL cells which express the adult globin genes. K562 and MEL cells are, however, transformed cells probably frozen at specific developmental stages; to what extent they resemble normal erythroid cells is disputable. We have used normal erythroid cells from the hamster yolk sac (Boussios et al, 1985) as transfection hosts for CAT plasmids containing the 1.9 Kb or the 0.8 Kb enhancer fragment. Our preliminary results (T. Boussios and D. Tuan, unpublished observations) indicate that the human erythroid enhancer, when coupled to the human embryonic ε or adult β globin promoter, is also functional in embryonic hamster erythroid cells, which express the embryonic as well as the adult b-like hamster globin genes.

Discussion: The above experiments suggest that in the human genome the enhancer element may be the primary sequence involved in transcriptional activation not only of the proximal embryonic ε-globin gene but also of the adult β globin gene at 53 to 54 Kb far downstream. Evidence in support of this suggestion is provided by the molecular defects of three reported cases of γδβ-thalassemias. In the Dutch (Van der ploeg et al, 1980; Vanin et al, 1983), the English (Curtin and Kan, 1988) and the Hispanic (Driscoll et al, 1987) γδβ thalassemias, the far upstream DNA spanning the enhancer element is deleted. The apparently normal β-globin genes in these γδβ-thalassemias are flanked by normal DNA 3' of the β-globin gene, and 5' of the β-globin gene for 2 Kb (Dutch) 25 Kb (English) and 50 Kb (Hispanic). Yet the in vivo expression of these β-globin genes is suppressed. In these cases of thalassemias, the enhancer sequence 3' of the Aγ globin gene and/or the enhancer 3' of the β-globin gene (Bodine and Ley, 1987; Kollias et al, 1987; Trudel and Constantini, 1988) are present; yet their presence does not seem to be sufficient to offset the effect of the common deletion of the far upstream DNA and the enhancer element in allowing full expression of the β-globin genes. The enhancer element underlying HSII thus appears to play a pivotal role in the transcriptional activation of the far downstream β-globin gene.

Further evidence suggesting a dominant role of the globin enhancer in transcriptional regulation of the adult β-globin gene is provided by transgenic experiments. In the transgenic mice, a human β globin transgene in a transchromosomal minilocus which is flanked on the 5' side by 22 Kb of DNA spanning erythroid specific major hypersensitive sites including HSII has recently been shown to be transcribed in mouse erythroid cells at a level equivalent to that of the endogenous mouse β globin gene (Grosveld et al, 1987). It is possible that the erythroid enhancer element in 0.8 Kb of DNA may constitute the sole sequence requirement for efficient transcription of a cis linked and integrated β-globin transgene in erythroid cells. The enhancer element may thus hold promise for ultimate gene therapy in human genetic disorders of erythroid cells such as the thalassemia syndromes and sickle cell anemia.

Acknowlegement: We thank Dr. M. Fordis for the enhancerless PβGLCAT plasmid. This work was supported in part by MIT intramural funds, by grants from Johnson and Johnson, and from the National Institutes of Health (DK 16272, HL 39948).

References

Bodine, D.M., and Ley, T. (1987) EMBO 6, 2997-3004.
Boussios, T., Condon, M., and Bertles, J. (1985) Proc. Natl. Acad. Sci. 82, 2794-2797.
Chada, K., Magram, J., and CostantiniF. (1986) Nature 319, 685-698.
Curtin, P., and Kan, Y.W. (1988) Blood 71, 766-770.
Driscoll, M.C., Dobkin, C., and Alter, B.P. (1987) Blood 70, 74a.
Dzierzak, E.A., Papayannopoulou, T., and Mulligan, R.C. (1988) Nature 331, 35-41.
Fordis, M., Nelson, N., Dean, A., Anagnou, N., Nienhuis, A., McCormick, M., Padmanabhan, R., and Howard, B. (1985) in Experimental Approaches for the Study of Hemoglobin Switching, A.R. Liss Inc. p. 281-292.
Forrester, W., Takegawa, S., Papayannopoulou, T., Stamatoyannopoulos, G., and Groudine, M. (1987) Nucl. Acids. Res. 15, 1015A.
Grosveld, F., van Assendelft, G.B., Greaves, D.R., and Kollias, G. (1987) Cell 51, 975-985.
Kollias, G., Hurst, J. deBoer, E., and Grosveld, F. (1987) Nucl. Acids. Res. 15, 5739-5748.
Kollias, G., Wrighton, N., Hurst, J., and Grosveld, F. (1986) Cell 46, 89-94.
Magram, J. Chada, K., and Costantini,F.(1985) Nature 315, 338-340.
Melton, D.A., Krieg, P.A., Rebagliati, M.R., Maniatis, T., and Green, M.R. (1984) Nucl. Acid Res. 12, 7035-7056.
Rutherford, T., and Nienhuis, A.W. (1987) Mol. and Cell. Biol. 7, 398-402.
Townes, T.M., Lingrel, J.B., Chen, Y. H., Brinster, R.L., and Palmiter, R.D. (1985) EMBO 4, 1715-1723.
Tuan, D., Abeliovich, A., Lee-Oldham, M., and Lee, D. (1987) In: Developmental Control of Globin Gene Expression, A.R. Liss, Inc., p. 211-220.
Tuan, D., Solomon, W.B., Li, Q.L., and London, I.M. (1985) Proc. Natl. Acad. Sci. 82, 6384-6388.
Trudel, M., and Costantini F. (1988) Genes and Development 1, 954-961.
Van der Ploeg, L. H. T., Lonings, A., Oort, M., Roos, D., Bernini, L., and Flavell, R.A. (1980) Nature 283, 637-641.
Vanin, E.F., Henthorn, P.S., Kioussis, D., Grosveld, F., and Smithies, O. (1983) Cell 35, 701-711.

Hemoglobin Switching, Part A: Transcriptional Regulation, pages 73–87
© 1989 Alan R. Liss, Inc.

REGULATORY PROTEIN ACTION IN THE NEIGHBORHOOD OF CHICKEN
GLOBIN GENES DURING DEVELOPMENT

G. Felsenfeld, T. Evans, P.D. Jackson, J. Knezetic,
C. Lewis, J. Nickol and M. Reitman

Laboratory of Molecular Biology, NIDDK, National
Institutes of Health, Bethesda, Maryland 20892.

INTRODUCTION

The globin family presents a gene system in which it
is possible to study both those regulatory mechanisms
responsible for developmental activation of an entire group
of related genes, and those which modulate the expression
of individual members of the group. We have carried out
such studies of the α- and β-globin gene families in
chicken. Our results provide examples of both kinds of
mechanisms.

The organization of the α- and β-globin domains (Engel
and Dodgson, 1978; Dodgson et al., 1981; Villeponteau and
Martinson, 1981) is shown in Figure 1. The four members of
the β family are the embryonic genes ρ and ϵ, the adult
gene β^A, and the hatching gene, β^H. The α family has three
members: the single embryonic gene π, and the two adult
genes α^A and α^D. All of the embryonic genes are expressed
only in cells of the primitive lineage, which are produced
until days 4-5 of embryonic development. The primitive
cells also express both 'adult' α genes, α^D and α^A. At
about day 5, the cells of the definitive lineage first
appear in the circulation. No embryonic genes are expressed
in these cells, but they continue to express α^D and α^A.
The β-family genes β^H and β^A are also active in definitive
cells, but β^H is turned off at about the time of hatching.
(See Bruns and Ingram, 1973).

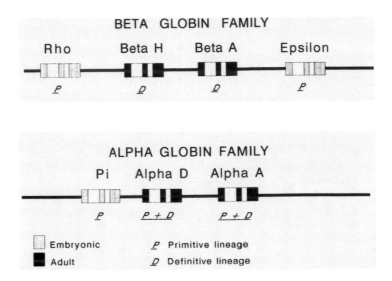

Fig. 1. The genomic arrangement and pattern of expression of the chicken globin gene family (see text).

The chicken is a particularly useful organism for the study of globin gene expression because erythroid cells at any stage of development can be obtained readily. Furthermore, the method we have devised (Hesse et al., 1986; Lieber et al., 1987) for introducing DNA into primary erythroid cells allows us to carry out transient expression studies with cells from embryos at each developmental stage.

RESULTS

The β-globin enhancer

In the first of these investigations (Hesse et al., 1986) we examined the function of β-globin regulatory elements. We used a plasmid carrying the β^A-globin promoter coupled to the gene for chloramphenicol acetyl transferase (CAT) as shown in Figure 2. Only low levels of CAT expression were observed when this plasmid was introduced into 9 day embryonic erythrocytes. Earlier reports (W. Schaffner, personal communication) had suggested that elements downstream of the gene might also play a role in expression; we therefore contructed plasmids in which various DNA segments from within the β^A-globin gene or its 3' flanking region were inserted 3' of the cat test gene. In this way we identified a DNA fragment (E in Fig. 2), starting about 110 bp 3' of the β^A-globin polyadenylation signal, which stimulates CAT expression nearly a hundred fold. The element functions when inserted either upstream or downstream of the promoter, and in either orientation, fulfilling the definition of an enhancer. The same enhancer has been identified independently by Choi and Engel (1986) and also described by Kretsovali et al. (1987); a related enhancer has been found 3' of the human β-globin gene (Trudel and Costantini, 1987; Behringer et al., 1987).

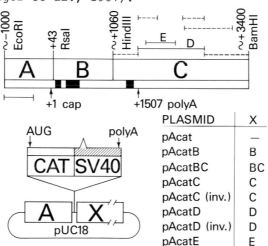

Fig. 2. Structure of globin-cat vectors. Shown above are the β^A-globin gene (exons in solid black) and the relative positions of the regions A-E used in the study. The globin promoter region (A) was inserted 5' of cat-SV40. The fragment X, chosen from among those listed at the lower right, was inserted just 3' of cat-SV40. Only fragments that include region E have enhancing activity in transient expression studies (Hesse et al., 1986).

Recent work has given us a detailed picture of individual elements that comprise this enhancer. Deletion analysis of the E fragment showed (Emerson et al., 1987) that all of

the stimulatory signals are contained in a 126 bp
subfragment (Fig. 3). DNase I footprinting revealed four
protected regions within this subfragment. In order to
assess the role of each of these elements in enhancer
function, a series of scanning mutants was synthesized,
spanning the entire enhancer domain (Reitman and
Felsenfeld, 1988). The effect of these mutations on CAT
activity is shown in Fig. 4. It is evident from the
mutations that footprint regions II and IV are the only
ones necessary for the full stimulatory effect of the
enhancer. Mutations of both regions I and III have no
effect on CAT activity; furthermore, constructions
containing three copies of either region II or region IV
are quite efficient enhancers in the absence of the other
elements.

This enhancer is approximately equidistant from the β^A
promoter in the 5' direction and the ϵ promoter in the 3'
direction.   Since no other positive regulatory elements
have been detected upstream of the ϵ gene or for a distance
of 6kb downstream (Nickol and Felsenfeld,1988), it seems
likely that this enhancer also serves to regulate
expression of the ϵ-globin gene in vivo. Studies in vitro
give results that are reasonably consistent with this idea:
The enhancer appears to be capable of acting on both
promoters in a fashion that partly mimics tissue
specificity (Nickol and Felsenfeld, 1988; Choi and Engel,
1988). Particularly dramatic effects are observed when the
enhancer and both genes are present on the same plasmid
(Choi and Engel, 1988).

BINDING DOMAINS OF THE BETA GLOBIN ENHANCER

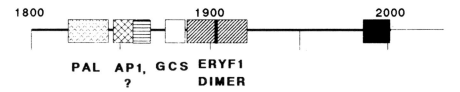

Fig. 3. The binding regions within the β-globin enhancer
were first detected by DNaseI footprint analysis (Emerson
et al., 1987). The four regions bind at least 5 distinct
factors, as shown. The solid black box at the right marks
a fifth footprint that lies outside the positive regulatory
domain.

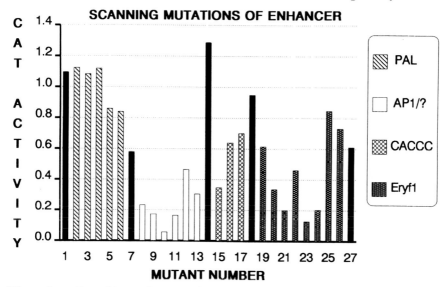

Fig. 4. Results of transient expression experiments in 10 day embryonic cells with scanning mutations (4 base transversions) of the β^A/ϵ enhancer. The relative efficiency of CAT expression is plotted as a function of the position of the mutation. (Reitman and Felsenfeld, 1988).

The enhancer contains binding sites for both ubiquitous and erythroid-specific factors. Footprint regions I and III contain sites for the Pal and GCS factors, respectively. These factors also bind to the β^A promoter, as discussed below. Region II appears to interact with two distinct proteins. The binding site of one of these is related to the sequence recognized by members of the c-jun/AP1 family (Angel et al., 1988); the other is not yet identified. Region IV binds two copies of a factor which we have named Eryf1. Footprint and gel retardation studies show that its binding activity is entirely confined to erythroid cells (Evans, Reitman and Felsenfeld, 1988).

The Eryf1 binding site is similar in sequence to the predominant binding site in the promoter of the α^D gene, which we had identified earlier (Kemper et al., 1987). We had also observed such a sequence in the promoter of the

ρ-globin gene, and shown by competition studies that the same factor bound to both the α^D and ρ-globin sites. The same methods, applied to the Eryf1 site, now reveal that Eryf1 and the earlier identified factor are identical (Fig. 5).

We surveyed the other members of the globin gene family for potential Eryf1 binding sites (Evans et al., 1988). Where necessary, we synthesized oligonucleotides containing the candidate sequences and measured binding directly by gel retardation assays. We find at least one strong Eryf1 binding site [with the core consensus sequence (A/T)GATA(A/G)] in either the enhancer or promoter of each family member (Table 1). We can also identify a strong Eryf1 site in the 3' enhancer of the human β-globin gene. Furthermore, the human erythroleukemia line, K562, contains factors that binds to this site (Fig. 5). The human factor NFE1 recently described by Grosveld and his collaborators appears to be one of these. We have purified Eryf1 to apparent homogeneity; it is an ~36 kD protein.

Table 1. Eryf1 binding sites are present in regulatory regions of all chicken globin genes. (From Evans et al., 1988)

Element	Position	Sequence
β-like Genes		
β^A/ϵ enhancer	+1891/+1906	gttgc AGATAA acatt
β^A/ϵ enhancer	+1920/+1905	agtct TGATAG caaaa
ρ promoter	-211/-196	cagca AGATAA gggct
β^H promoter	-157/-172*	aggaa AGATAG caaat
α-like Genes		
α^D promoter	-67/-52	ggata AGATAA ggccg
π promoter	-176/-161*	caagg AGATAA gggtc
α^A enhancer	+1417/+1402	ggccg TGATAA gagcc
α^A enhancer	+1459/+1474	gtggc TGATAA agagc
α^A enhancer	+1518/+1533	gcagc AGATAG cctcg

Shown are some of the DNA sequences of chicken globin regulatory regions that are known to bind factors from erythroid cells. All binding sites contain the motif (A/T)GATA(A/G), and most likely represent interactions with Eryf1. Only the strand that contains the GATA motif is

shown, relative to the position of the mRNA start site, or
the ATG translational initiation codon (*).

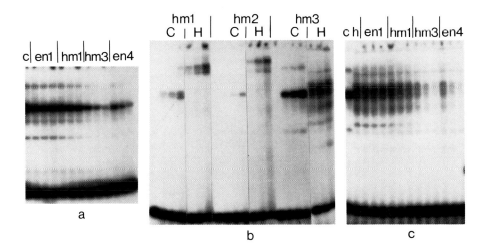

Fig. 5. Gel retardation experiments showing that factors in
human erythroid cells and chicken Eryf1 crossreact with
common sequences in the human β-globin 3' enhancer, the
chicken β^A/ϵ enhancer, and the chicken ρ-globin promoter.
(a) An oligonucleotide containing an Eryf1 site from the
chicken ρ-globin promoter was used as labelled probe.
Adult chicken erythrocyte nuclear extracts were used alone
(lane c) or in the presence of increasing amounts of
unlabelled oligomers containing DNA sequence from the Pal
binding site of the adult β enhancer (en1), the Eryf1
binding site from the same enhancer (en4), or potential
Eryf1 sites from the human β enhancer (hm1, hm3).   (b)
Three potential Eryf1 sites from the human β enhancer
(hm1,2,3) were labelled and incubated with chicken (C) or
human K562 (H) extracts.   The hm3 sequence binds most
strongly, and in a manner similar to chicken Eryf1.   The
pattern with human extracts is however more complex.   (c)
Oligomer hm3 was incubated with chicken extract (lane c) or

K562 extract (all other lanes), either in the absence (lane h) or presence of competing en1, hm3, or en4. (From Evans et al., 1988).

The β^A-globin promoter

Studies of chromatin structure carried out in our laboratory some years ago (McGhee et al., 1981) focused our interest on the β^A-globin promoter, which we found to occupy, in definitive erythroid cells, a nucleosome-free region within which were bound a number of non-histone proteins. Our present state of knowledge of the structure of this domain is summarized in Fig. 5. Like many other promoters, the β^A-globin promoter contains a large number of binding sites for sequence-specific proteins. These sites have been identified both by footprinting (Emerson et al., 1985) and by gel retardation assays (Lewis et al., 1988). Some of these bind ubiquitous regulatory factors: In addition to the usual CAAT and TATA sequences, the β^A-globin promoter contains one weak and one strong (Evans, DeChiara and Efstratiadis 1988; Lewis et al., 1988) binding site for Sp1. Binding sites are also found for factors that are more restricted in the range of tissues in which they are expressed or functional.

ADULT BETA GLOBIN PROMOTER

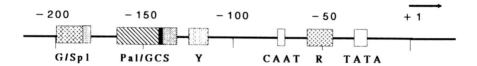

Fig. 6. Binding sites of the β^A-globin promoter. G = G_{16}, which binds BGP1; GCS = globin consensus sequence (CACCC); Pal = Palindrome; Y and R are polypyrimidine and polypurine sites that are respectively weak and strong Sp1 binding sites. (See Lewis et al., 1988; Jackson, 1987; Evans, DeChiara and Efstratiadis, 1988).

At the 5' end of the domain is a string of 16 G residues, terminating in the sequence GGGCGG, a potential

Sp1 binding site. However Sp1 binds only weakly to this region (Lewis et al., 1988). The principal factor binding to the poly dG sequence is a protein which we have named BGP1. This protein requires Zn^{++} for its binding activity; a monoclonal antibody raised to BGP1 cross-reacts weakly with Sp1. These results suggest that BGP1 may be a 'zinc finger' protein. Transient expression in primary erythrocytes of mutants in which the poly dG sequence is deleted have not yet identified a function for this site; we have suggested that BGP1 may be involved in the organization of chromatin structure at the hypersensitive domain (Lewis et al., 1988).

The two elements immediately downstream of the G string are the palindrome sequence (Pal) and the CACCC sequence [which we refer to as the globin consensus sequence, GCS, since it is common to most adult β-globin promoters (Dierks et al., 1983)]. Both of these sequences are found also in the enhancer, as described above. Transient expression studies with cat constructions in which Pal and GCS are successively deleted from the promoter show that Pal acts as a negative regulatory element, and GCS as a positive element. The regulatory properties of GCS have been reported in other systems (Dierks et al., 1983; Myers et al., 1986).

The protein binding to Pal is a member of the nuclear factor 1 (NF1) family, but it appears to be unique to erythroid cells (C. Lewis, unpublished data). We have measured the abundance of the Pal protein both by gel retardation and footprint assays of nuclear extracts, and we have also used an intranuclear footprinting method (Jackson and Felsenfeld, 1985) to determine the extent of occupancy of the Pal site within the nucleus (Jackson, 1987; unpublished data). We find that there is little or no Pal factor present (or bound) in 5 day embryonic cells, and only a small amount at day 9. Pal factor becomes abundant and the site becomes fully occupied only late in development (day 15 to adult) when β^A-globin synthesis is decreasing. These results, combined with the deletion data, suggest that Pal helps to shut down transcription from the β^A-globin gene.

Within the nucleus, the binding of the GCS protein to its site diminishes with embryonic age as Pal binding increases. Competition binding studies (unpublished data) show that the decrease is not the result of displacement of GCS protein by Pal protein. In any case, reduced binding to the GCS is likely also to have the effect of decreasing β^A-globin gene transcription.

The α-globin regulatory regions

A survey of the adult α-globin domain reveals, in addition to the α^A and α^D promoters, a single enhancer in the 3′ flanking region of the α^A gene (Knezetic and Felsenfeld, 1989). The only regulatory elements we have been able to detect within this enhancer are three Eryf1 binding sites (Fig. 7). The α^D gene has no enhancer; as discussed earlier its promoter consists principally of a single Eryf1 binding site. The relative simplicity of the regulatory regions of these two genes contrasts with the complexity of regulation of β^A-globin. Perhaps this reflects the difference in developmental programs for the adult α and β^A genes: The two adult α genes are expressed in both primitive and definitive cells, while β^A is not expressed in primitive cells. In principle, the α genes thus require only the positive signal of Eryf1 binding, while the β^A gene requires additional modulation.

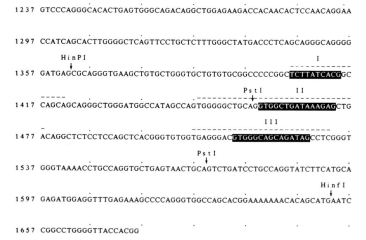

Fig. 7. DNase I footprints (dark blocks) of 9-day embryonic erythrocyte extracts on the 3′ flanking region of the α^A globin gene, within the enhancer. The three footprinted regions are all Eryf1 binding sites. From Knezetic and Felsenfeld, 1989.

DISCUSSION

There is a hierarchy of mechanisms regulating expression of the globin gene family. At the highest level, the dominant control region or locus activation region (Grosveld et al., 1987; Forrester et al., 1987; Tuan et al., 1985) probably acts in a tissue-specific fashion on the chromatin structure of an entire domain. It should soon be clear whether this region employs a unique set of regulatory proteins for its function, or shares at least some of the same regulatory factors (for example Eryf1) that bind near the genes of the cluster.

At the next level of organization, Eryf1 appears to be a major erythroid-specific regulatory protein. We have found at least one Eryf1 binding site in one of the regulatory domains - either promoter or enhancer - of every member of the α- and β-globin families. Furthermore, potential Eryf1 sites are found in the histone H5 enhancer, and in the human β-globin enhancer and promoter (Evans, Reitman and Felsenfeld, 1988; Wall et al., 1988), as well as within a DNase I hypersensitive site of the mouse β-globin gene (Galson and Housman, 1988). We suggest that Eryf1 is one of the principal determinants of erythroid-specific expression.

It is evident that since individual members of the globin family have distinct repertoires of developmentally regulated expression, a regulatory factor like Eryf1 cannot be sufficient. Our results suggest that while the Eryf1 site within the enhancer of the adult β-globin gene confers general erythroid-specific expression, this gene's promoter contains binding sites for several other factors, some erythroid-specific and others ubiquitous, that must serve to confine its expression to active definitive cells. The Pal factor, for example, is an erythroid-specific protein that increases in abundance rather late in definitive cell development, when globin expression is being shut down. On the other hand, intranuclear binding to the positive globin consensus sequence (GCS) decreases late in development. We have shown that the Pal binding site is a negative regulatory element; the GCS is a positive element. Both presumably function in the orderly shutdown of β^A-globin expression.

In contrast, the α^A and α^D-globin genes seem to have much less complex promoters and enhancers, dominated by Eryf1 sites. Eryf1 alone, or in combination with a small number of other factors, might well be sufficient for the control of these genes, which are expressed at every stage of erythroid development.

There is reason to be optimistic that continued analysis of the individual regulatory factors of the globin family will lead to a reasonably complete picture of the reactions that control developmental switching. As in many other cases, these genes are regulated by combinations of transcriptional factors, many of which are ubiquitous. However some (such as Pal) are specific to certain stages of erythroid development, and one, Eryf1, is present at all stages, but is completely unique to erythroid cells. It will be interesting to determine the mechanism of action of these specific components, and particularly the mechanism by which their own expression is regulated at the earliest times of erythroid development.

REFERENCES

Angel P, Allegretto EA, Okino S, Hattor K, Boyle WJ, Hunter T, Karin M (1988). The jun oncogene encodes a sequence specific trans-activator similar to AP-1. Nature 332: 166-171.

Behringer RR, Hammer RE, Brinster RE, Palmiter RD, Townes TM (1987). Two 3' sequences direct erythroid-specific expression of human β-globin genes in transgenic mice. Proc Natl Acad Sci USA 84:7056-7060.

Bruns GA, Ingram VM (1973). The erythroid cells and haemoglobins of the chick embryo. Phil Trans Roy Soc Lond (Biol) 266:255-305.

Choi O-R, Engel JD (1986). A 3' enhancer is required for temporal and tissue-specific transcriptional activation of the chicken adult β-globin gene. Nature 323:731-734.

Choi O-R, Engel JD (1988). Developmental regulation of β-globin gene switching. Cell 55:17-26.

Dierks P, van Ooyen A, Cochran MD, Dobkin C, Reiser J, Weissmann C (1983). Three regions upstream from the cap site are required for efficient and accurate transcription of the rabbit β-globin gene in mouse 3T6 cells. Cell 32:695-706.

Dodgson JB, McCune KC, Rusling DJ, Krust A, Engel JD (1981). Adult chicken alpha-globin genes alpha A and alpha D: no anemic shock alpha-globin exists in domestic chickens. Proc Natl Acad Sci USA 78:5998-6002.

Emerson BM, Nickol JM, Jackson PD, Felsenfeld G (1987). Analysis of the tissue specific enhancer at the 3' end of the chicken adult β-globin gene. Proc Natl Acad Sci USA 84:4786-4790.

Emerson BM, Lewis CD, Felsenfeld G (1985). Interaction of specific nuclear factors with the nuclease-hyper-sensitive region of the chicken adult β-globin gene: Nature of the binding domain. Cell 41:21-30.

Emerson BM, Felsenfeld G (1984). Specific factor conferring nuclease hypersensitivity at the 5' end of the chicken adult beta globin gene. Proc Natl Acad Sci USA 81:95-99.

Engel JD, Dodgson JB (1978). Analysis of the adult and embryonic chicken globin genes in chromosomal DNA. J Biol Chem 253:8239-46.

Evans T, DeChiara T, Efstratiadis A (1988). A promoter of the rat insulin like growth factor II gene consists of minimal control elements. J Mol Biol 199:61-81.

Evans T, Reitman M, Felsenfeld G (1988). An erythrocyte-specific DNA binding factor recognizes a regulatory protein common to all chicken genes. Proc Natl Acad Sci USA 85:5976-5980.

Forrester WC, Takegawa S, Papayannopoulou T, Stamatayannopoulos G, Groudine M (1987). Evidence for a locus activation region: the formation of developmentally stable hypersensitive sites in globin-expressing hybrids. Nucleic Acids Res. 15:10159-77.

Galson DL, Housman DE (1988). Detection of two tissue-specific DNA binding proteins with affinity for sites in the mouse β-globin Intervening Sequence 2. Mol Cell Biol 8:381-392.

Grosveld F, Blom van Assendelft G, Greaves DR, Kollias G (1987). Position-independent, high-level expression of the human β-globin gene in transgenic mice. Cell 51: 975-985.

Hesse JE, Nickol JM, Lieber MR, Felsenfeld G (1986). Regulated gene expression in transfected primary chicken erythrocytes. Proc Natl Acad Sci USA 83:4312-4316.

Jackson P D (1987). Doctoral Thesis, Johns Hopkins University.

Jackson PD, Felsenfeld G (1985). A method for mapping intranuclear protein-DNA interactions and its application to a nuclease hyper-sensitive site. Proc Natl Acad Sci USA 82:2296-2300.

Kemper B, Jackson PD, Felsenfeld G (1987). Protein binding sites within the 5' DNase I hypersensitive region of the chicken αD-globin gene. Mol Cell Biol 7:2059-2069.

Knezetic J, Felsenfeld G (1989). Identification and Characterization of a Chicken α-Globin Enhancer. Molecular and Cellular Biology, in press.

Kretsovali A, Muller MM, Weber F, Marcaud L, Farache G., Schreiber E, Schaffner W, Scherrer K. (1987). A transcriptional enhancer located between adult beta-globin and embryonic epsilon-globin genes in chicken and duck. Gene 58:167-175.

Lewis CD, Clark SP, Felsenfeld G, Gould H (1988). An erythrocyte-specific protein that binds to the poly(dG) region of the chicken β-globin gene promoter. Genes and Development 2:863-873.

Lieber MR, Hesse JE, Nickol JM, Felsenfeld G (1987). The mechanism of osmotic transfection of avian embryonic erythrocytes:Analysis of a system for studying developmental gene expression. J Cell Biol 105: 1055-1065.

McGhee JD, Wood WI, Dolan M, Engel JD, Felsenfeld G (1981). A 200 base pair region at the 5' end of the chicken adult β-globin gene is accessible to nuclease digestion. Cell 27:45-55.

Myers RM, Tilly K, Maniatis T (1986). Fine structure genetic   analysis of the β-globin promoter. Science 232:613-618.

Nickol JM, Felsenfeld G (1988). Bidirectional control of the chicken β- and ε-globin genes by a shared enhancer. Proc Natl Acad Sci USA 85:2548-2552.

Reitman M, Felsenfeld G (1988). Mutational analysis of the chicken β-globin enhancer reveals two positive-acting domains. Proc Natl Acad Sci USA 85: 6267-6271.

Trudel M, Costantini F (1987). A 3' enhancer contributes to the stage-specific expression of the human β-globin gene. Genes and Dev 1:954-961.

Tuan D, Solomon W, Li Q, London I (1985). The 'β-like-globin' gene domain in human erythroid cells. Proc Natl Acad Sci USA 82:6384-6388.

Villeponteau B, Martinson H (1981). Isolation and characterization of the complete chicken β-globin gene region:frequent deletion of the adult β-globin genes in λ. Nucleic Acids Res 9:3731-3746.

Wall L, deBoer E, Grosveld F (1988). The human β-globin gene 3' enhancer contains multiple binding sites for an erythroid-specific protein. Genes and Dev 2:1089-1100.

Hemoglobin Switching, Part A: Transcriptional Regulation, pages 89–103
© 1989 Alan R. Liss, Inc.

GENETICS AND BIOCHEMISTRY OF THE EMBRYONIC TO ADULT SWITCH IN THE CHICKEN ε- AND ß-GLOBIN GENES

James Douglas Engel, Ok-Ryun B. Choi, Debra J. Endean, Kevin P. Foley, James L. Gallarda and Zhuoying Yang

Department of Biochemistry, Molecular Biology and Cell Biology, Northwestern University, Evanston, Il. 60208

Erythroid cell transcriptional regulation is usually characterized by focusing primarily on the two most prominent features of hematopoiesis; the control of globin gene expression in chickens provides a typical example. First, globin genes are transcribed only in cells of the erythroid lineage (1); and second, different genes are expressed at distinct developmental stages during ontogeny (2). The tissue specificity for adult ß-globin gene transcription is governed by the activity of a cellular enhancer sequence in chickens (3, 4) and by enhancer (and probably other distal) control sequences in the analogous human gene (5-9 and other papers in this volume). The second process (how genes "switch" hemoglobin subunit isotypes during embryonic development) is currently the least well understood of these two most prominent phenomena.

GENETICS OF THE ß/ε-GLOBIN DEVELOPMENTAL SWITCH

The chicken ß-globin gene cluster is arranged such that the enhancer is located almost precisely equidistant from the 5' adult ß-globin and 3' embryonic ε-globin gene promoters (Fig. 1). As we had originally speculated (3), we and others were recently able to demonstrate that this enhancer regulates both genes (10, 11). In a series of transfection experiments where marked ß- or ε-globin gene constructs (called ß* and ε*) were transfected into primary transformants of primitive or definitive erythroid precursor cells, we showed that while the adult ß-globin gene is cell stage-specifically regulated in a developmentally appropriate manner, the embryonic ε-globin gene is expressed at

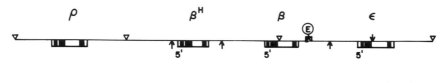

Figure 1. Organization of the Chicken ß-globin Genes.
The four genes of the ß-globin locus are shown relative to
the restriction map and transcription units for each gene
within the locus (12, 13). The position of the ß-globin
enhancer is depicted on the line as a circled "E" (3, 4).

equivalent levels in either primitive (embryonic) or definitive (adult) red cells (10; Fig. 2). These data led to the
conclusions that while the adult ß-globin gene contains
sequences which are both necessary and sufficient for it's
tissue- and stage-specific regulation, the embryonic ε-globin
gene is not properly regulated (with respect to stage
specificity) in the constructs examined (10).

A key experiment in further studies showed that when the
embryonic ε- and adult ß-globin genes were transfected into
definitive erythroid cells on the same plasmid in their
normal genomic configuration, proper stage-specific regulation was achieved for both genes [the embryonic ε-globin gene
was now transcriptionally silent in definitive cells; Fig.
3 and (10)]. These data implied that other DNA sequences
(within the adult ß-globin gene locus) are required for
proper stage-specific ε-globin gene regulation, and that such
sequences were not present on the original ε-globin plasmids
tested.

Several experiments were then performed to ask where the
ε-globin effector sequences were located within the adult ß-
globin locus. Based on predictions from DNA sequence
analysis of the ß-globin gene family (12, 13), we examined
the possiblity that the cis-effector sequence(s) responsible
for ε-globin quiesense in definitive cells might lie within
the adult ß-globin gene promoter. After examination of a
series of promoter substitution and deletion mutants, we con-
cluded that the ε-globin cis-effector sequences were located
within a small segment of the adult ß-globin gene promoter,
between nucleotides -112 and -11 relative to the mRNA capsite
(10, 12). That same segment of DNA was shown in other
experiments to contain a genetic element which allows the

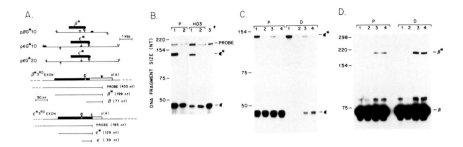

Figure 2. Transfection of Marked Globin Genes into AEV-transformed Primitive or Definitive Erythroid Progenitors. ε*-plasmid constructs containing or lacking the ß-globin enhancer (the black box in panel A.) were transfected into primitive (P) or definitive (D) erythroid precursor cells and tested for mRNA accumulation after differentiation induction using S1 nuclease protection of the marked gene probes (A.; 10). Panel B. shows the results of ε (and ε*) mRNA accumulation after transfection into P or HD3 (15) cells followed by differentiation induction; lanes 1 = ε*-globin containing the ß-globin enhancer, lanes 2 = ε*-globin lacking the enhancer. C. and D. show the results of transfection of the same constructs into P and D cells either without (lanes 1, 2) or with (lanes 3, 4) cotransfected pßG*10 (even numbered lanes = enhancer (-) ε*-globin genes; odd numbered lanes = ε*-globin genes with the enhancer; 3, 10). Panel C. represents the analysis of ε*-globin accumulation; panel D. represents the analysis of ß*-globin mRNA accumulation.

adult ß-globin gene to be preferentially transcribed (when compared to ε-globin) in definitive erythroid cells (10). We termed this new genetic element the SSE, for developmental Stage Selector Element. Clearly, two possibilities for the activities present in the adult ß-globin promoter region exist: the genetic elements responsible for preferential ß-globin transcription and for ε-globin cis-suppression in definitive erythroid cells might be one and the same, or two different elements within the ß-globin promoter confer these distinct regulatory properties.

In order to determine if ε-globin suppression in definitive erythroid cells is due to direct negative regulation by the ß-globin promoter or if the adult ß-globin promoter somehow indirectly effects ε-globin transcription, we

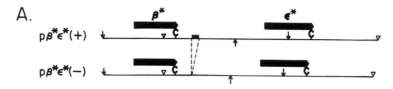

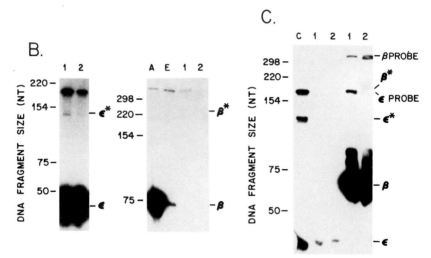

Figure 3. Transfection of *cis*-linked ε- and ß-globin Genes. The ε* and ß* globin genes were placed on one plasmid either with (+) or without (-) the ß-globin enhancer (A.), and were transfected into circulating 4.5 da embryonic erythroblasts (B.; 4) or D cells (C.; 10) and tested for expression of both genes. Lanes 1 = (+) construct; lanes 2 = (-) construct. A = adult (definitive) RBC RNA; E = 4.5 da embryonic (primitive) RBC RNA; C = a sample of transfected cell RNA (identical to that shown in Fig. 2C, lane P1).

prepared a construct in which the enhancer was duplicated. The logic used in design of this experiment suggested that if the ß-globin promoter exerts a direct negative effect on ε-globin transcription, duplication of the enhancer should not effect this presumptive negative regulation. Alternatively, if the adult ß- and embryonic ε-globin gene promoters simply compete for enhancer activity, duplication of the enhancer should lead to transcriptional activation of both genes in definitive erythroid cells.

The results of transfection of the duplicated enhancer plasmid, containing both genes, into definitive cells showed that both genes are active (10). Thus, ε-globin suppression is due to an indirect mechanism, whereby ß-globin promoter sequences compete for all of the activity of the ß-globin enhancer in definitive erythroid cells, to the exclusion of the ε-globin promoter's ability to do so.

On the basis of these data, we proposed a model wherein the choice between promoters was directed by the ß-globin enhancer (10; Fig. 4). This model suggests that the *trans*-acting factors which interact with the ß- and ε-globin promoters and the ß-globin enhancer also interact with one another to form a stable transcription complex through direct protein:protein interactions. The choice the enhancer makes between which promoter is to be utilized is presumed to be mediated by specific *trans*-acting factors (predicted to be differentially expressed at different developmental stages) and which bind to either the enhancer and/or the ß- and ε-globin promoters (or possibly to all three control sequences). This experimental prediction thus led to the examination of biochemical extracts of erythroid cells in order to ask which nuclear factors (that bind to the ß- and ε-globin promoters and to the enhancer *in vitro*) might be present in cells of different developmental stages.

BIOCHEMISTRY OF THE ß/ε-GLOBIN DEVELOPMENTAL SWITCH

We initiated biochemical studies based on the protocol for extract preparation developed by Emerson *et al*. (14). In this procedure, whole cell extracts are prepared by a final step of binding to, and elution from, double-stranded DNA cellulose. These basic extracts (as distinct from crude whole cell or nuclear extracts) are thus enriched in nucleic acid binding proteins. The relative concentration (/affinity) of individual proteins which bind to presumptive regulatory sites in DNA is then scored by DNaseI footprinting.

We initially set out to examine differences in proteins which bind to the enhancer, and their relative abundance in extracts prepared from primitive and definitive erythroid cells. In addition, we hoped to be able to discern whether

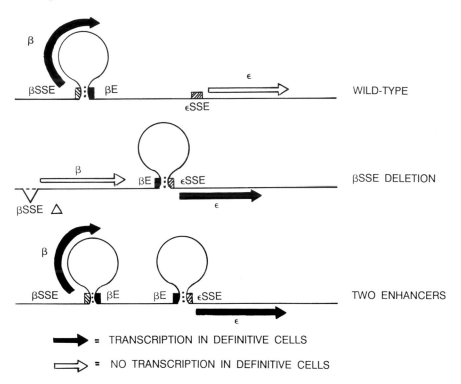

Figure 4. A Model for Promoter Competition for the ß-globin Enhancer. Analysis of adult ß-globin gene promoter deletion and substitution experiments in *cis* to unmodified ε-globin genes (10) suggests that a stable transcription complex is formed between the ß-globin enhancer and either the ß- or ε-globin promoters. The stage selector element (see text), if deleted, leads to ε-globin accumulation (10), whereas if the enhancer is duplicated, both genes are transcribed (10).

or not the binding of either constitutive or tissue-specific proteins varied, not only by developmental age, but also by the stage of erythroid cell development. To that end, we also prepared basic extracts from HD3 cells (a definitive avian erythroid precursor cell line, transformed by a ts mutant of AEV, arrested at approximately the CFU-E stage; 15). We anticipated that if different proteins in the various erythroid cell extracts were present at different concentrations, we might be able to construct logical arguments as to which of the factors might be important in regulating the switch.

Somewhat to our surprise, of the five different regions of the ß-globin enhancer which were originally defined as binding sites for proteins within definitive erythroid cells (16), extracts prepared from primitive, definitive and HD3 cells all showed much the same pattern of protection from DNaseI digestion *in vitro* (Fig. 5). [While these data are at variance with the original report showing *in vitro* DNase I footprints in the enhancer, we are not certain as to the source of these descrepancies; 16]. The only clear difference in the enhancer region between the various extracts is the presense or absense of ßE-P1*. Clearly, only one of two possibilities could then be thought to be responsible for the differential activity of the enhancer: either the identical enhancer binding proteins are present in each extract preparation (and consequently, differential *trans*-acting factor binding to the enhancer is not a direct regulatory step in the developmental decision as to whether the ß- or ε-globin gene is to be transcribed), or different enhancer binding proteins (with highly homologous binding sites) are activated at different developmental stages, and the stage-specific binding of these factors is responsible for the differential activity of the two genes. In support of the latter hypothesis, small differences in the pattern of DNaseI footprints are clearly evident in comparison of the results using the three different erythroid cell extract preparations (Fig. 5).

Since only few differences were visualized in the binding of various erythroid cell extract proteins to the ß-globin enhancer, we also examined the binding properties of factors from these same extracts to the ß-globin promoter region. In contrast to the results observed from examination of enhancer-binding factors, here we see evidence for moredramatic differences in the relative concentration of particular *trans*-acting factors from different stages of erythroid cells. As shown in Figure 6, four footprints are visualized in the -11 to -112 region (the segment of the adult ß-globin promoter containing the SSE) in definitive cell extracts. These likely correspond to the binding of "TATAA" factor, TFII-D (17; ßP-P1) at position -20 to -30,

---

*We refer throughout this paper to the DNA site recognized by a particular protein as "F" (e.g. ß-globin enhancer footprint 3 is abbreviated ßE-F3) while the protein giving rise to that same footprint is designated "P" (e.g. ßE-P3).

Figure 5. <u>DNAse I Footprinting of the ß-globin Enhancer.</u> Basic extracts were prepared from definitive (D), primitive (P) and immature definitive (HD3) cells. The volume of extracts shown were then bound to a 500 bp, singly end-labeled PvuI/BstNI fragment (16) containing the ß-globin enhancer, digested briefly with 1.6 µg/ml DNase I, phenol extracted, precipitated and finally visualized by electrophoresis on a sequencing gel. The designated footprints are given the same names as originally designated (16), with slight changes in nomenclature.

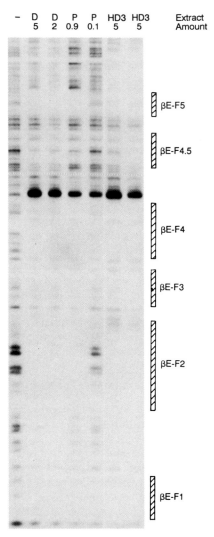

an unknown factor to position -35 to -48 (ßP-P2), CTF to position -65 to -80(18; ßP-P3) and transcription factor AP-2 to position -95 to -105 (19, 20; ßP-P4).

In Figure 6A, we compare the footprinting of basic extracts prepared from mature definitive cells and immature (transcriptionally active) primitive cells. The primitive cell extracts appear to have a much lower relative concentration of the proteins giving rise to ßP-F2 and -F3 in comparison to the concentrations of TFII-D and AP-2 (ßP-P1 and ßP-P4, respectively). The absense of ßP-P2 and ßP-P3 in primitive cells implies a causal link to the preferential activity of ε-globin gene transcription in primitive erythrocytes.

Figure 6B shows a comparison between mature and immature cells of the definitive erythroid lineage. HD3 cell extracts are representative of approximately the CFU-E stage of erythroid cellular maturation (15), a stage at which ß-globin is not yet expressed, but which is poised for overt ß-globin transcription (21). These cells differ in the promoter region from mature definitive erythroid cell extracts in only one respect; ßP-P2 is virtually absent in extracts prepared from HD3 cells. The strong inference from this *in vitro* experiment is that the only component required for adult ß-globin transcription which is missing in immature definitive (HD3) cells is ßP-P2.

"DUELLING LOLLIPOPS" AS A MODEL FOR DEVELOPMENTAL REGULATION

In attempting to derive a coherent, internally consistent explanation for the mechanism of the developmental switch in embryonic ε- to adult ß-globin gene transcriptional activation, we have used both genetic and biochemical observations. We have attempted to correlate these observations with DNA sequence analysis of the genes and to phenotypic characteristics expressed in developmentally different types of red cells, to explain in molecular detail how the switch might work.

In exploring the model shown in Figure 4 both more precisely and in much greater structural detail, one finds that if the ß-globin promoter and enhancer are physically aligned, there is a remarkable apposition of the binding sites for each of the proteins detected in definitive erythroid cell basic extracts; in essense, each protein binding site in the promoter can be directly aligned with a footprinting site in the enhancer. Since the ß- and ε-globin gene promoters are highly homologous (see below), it seems unlikely that this spacing of binding sites is purely coincidental. If we now superimpose the results of DNaseI footprinting of the promoter and enhancer sequences using the results accumulated in the biochemical studies (which show that different factors are found in different stages of erythroid cells), we find a remarkably simple correlation

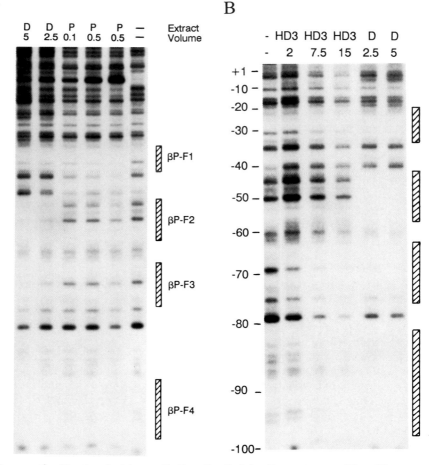

Figure 6. Footprinting of the ß-globin Promoter. The figure
depicts the DNase I footprints detected in the ß-globin promoter region using either definitive (D), primitive (P) or
immature definitive (HD3) cell basic extracts as in Fig. 5.

to observations between protein binding and the phenotype
of the various red cells examined here (Fig. 7). Because of
obvious structural similarities, we refer to this model
explaining the developmental choice of the enhancer between
the two promoters as "duelling lollipops".

In primitive erythroid cells, one finds that in the ß-
globin promoter region, only ßP-P1 and ßP-P4 are present,
while in the enhancer, all the proteins (with the exception

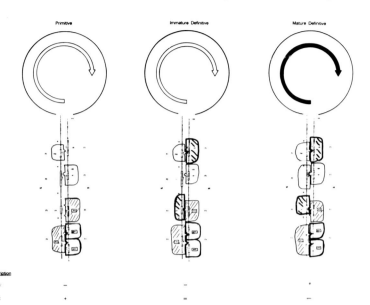

<u>Figure 7.</u> <u>Transcription "Lollipops" for the Chicken Adult ß-globin Gene.</u> The diagram shows the presumptive transcription complex formed between the ß-globin promoter and enhancer. In primitive RBC (left), only one favorable protein:protein interaction occurs; consequently, the ß-globin lollipop is unstable and ß-globin is not transcribed (open arrow). A lollipop is predicted to be stable in primitive cells only when the enhancer interacts with the embryonic ε-globin promoter (not shown), or in definitive cells when the ß-globin promoter he SSE) is deleted (10).   In immature definitive RBC (HD3) cells (middle), several productive protein:protein interactions take place which stabilize this stem (relative to the alternate one formed between the enhancer and the ε-globin promoter) and yet the adult gene is still not transcribed due to the absense of ßP-P2. However, in this cell ε-globin also cannot be transcribed since the ß-promoter/ß-enhancer interaction is preferred. In mature definitive cells (right), ßP-P2 is present, and transcription of the adult ß-globin gene progresses (filled arrow).

of ßE-P1) are present (ref. 14 and Fig. 5).   Under the assumption that stable transcription complex formation is dependent on multiple, weak protein:protein interactions between the enhancer- and promoter-binding *trans*-acting factors, we would then deduce that such a stable complex between the enhancer and the adult ß-globin promoter would not be formed in primitive erythroid cells [since the only contacts which could be formed between promoter and enhancer proteins would be between AP-2 in the promoter (ßP-P4) and NF-E1 in the enhancer (ßE-P4)].

In immature definitive cells (as represented by HD3 cell extracts), stable complex formation is now achieved by the multiple protein:protein contacts in the enhancer and adult gene promoter regions, and yet the gene is still not transcriptionally active, presumably due to the absense of ßP-P2 (Figure 6B). However, the formation of a stable adult ß-globin gene transcription complex also precludes ε-globin transcription, and consequently both genes are silent in undifferentiated HD3 cells. In mature definitive cells (or in HD3 cells induced for erythroid differentiation) ßP-P2 activity is revealed, thereby allowing ß-globin transcription.

Several explicit predictions of this model for ß-/ε-globin gene switching can now be formulated. First, one would anticipate that the DNA sequence represented by ßP-F2 would differ in the embryonic globin gene promoter. In comparing the DNA sequences of the adult and both embryonic ß-type globin genes, one in fact finds that the TFII-D, CTF and AP-2 binding sites (ßP-F1, -F3 and -F4, respectively) are all still present at the same relative position in the ε- and ρ-globin gene promoters; the only binding site which is different in the comparison is ßP-F2 [Fig. 8; (12, 13)]. This difference is manifest by the generation of a perfect AP-2 binding site at the same relative position of the promoter in the two embryonic genes, and (as shown in Figs. 5 and 6) transcription factor AP-2 (ßP-P4 or ßE-P3) is clearly present in both primitive and immature definitive erythroid cells. Thus we anticipate that while the ß-globin gene promoter would not be able to form a stable complex with the enhancer in primitive cells, the ε-globin promoter should be able to do so.

One puzzling aspect of this model is why CTF (ßP-P3) binding is not visualized in footprints using primitive cell extracts in examination of the ß-globin gene promoter. Since the identical core CCAAT sequence is present in both the ε- and ß-globin genes (12, 13), two possibilities might be suggested as to why CTF from primitive erythroid cells cannot

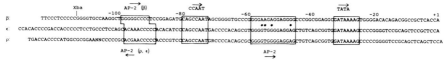

Figure 8. Sequence Comparison of the Adult (ß) and Embryonic
(ε and ρ) Globin Gene Promoters. DNA sequence data are taken
from references 12 and 13. Boxed sequences correspond to
either actual (ß-globin; Fig. 6) or theoretical (ε- or ρ-
globin) footprints. DNA sequence differences between the
promoters within the footprinted regions are indicated by
dots.

footprint the adult ß-globin gene promoter: either the factor
produced in primitive erythroid cells is different than that
found in definitive cells, or the context of CTF binding
(either in adjacent DNA sequences or in adjacent, closely
associated protein contact points) differs in the two cell
types. Since CTF in mammalian cells is actually a family of
proteins (18), both speculations appear to be viable possibilities given the extent of our present knowledge. We are
currently addressing this question by examination of footprints within the ε-globin gene promoter using these same
erythroid cell extracts.

There are also very explicit genetic predictions from
the model shown in Fig. 7. First, alteration of ßP-F2 to a
random DNA sequence should allow no ß-globin mRNA accumulation from an otherwise normal gene, even in differentiated
HD3 cells, since ßP-P2 is assumed to be the only factor
missing in the complex. Second, alteration of ßP-F2 to an
AP-2 binding site might be expected to allow ß*-globin gene
transcription in undifferentiated HD3 cells (since the model
suggests that the only factor missing in these cells is ßP-
P2, while transcription factor AP-2 is present in all
erythroid cell stages examined; e.g. Fig. 6). Both of these
predictions are being tested at the present time.

Another genetic prediction which follows from this model
is that the entity defined as the SSE (10) is, in reality,
a product of two separate biochemical processes: preferential
ß-globin promoter utilization in definitive erythroid cells
is due to the presense of ßP-P2 in those cells, while ε-
globin "suppression" by the ß-globin promoter is predicted
to be the result of the formation of a stable stem structure

between the ß-globin promoter factors ßP-P1, -P3 and -P4 with the enhancer factors ßE-P1, -P3 and -P4.

Central questions remain regarding the nature of the switch: what is the identity of the factor ßP-P2 (which apparently plays such a pivotal role in adult ß-globin gene transcription), and how is the activity of this protein regulated? Data from other laboratories (see papers by Felsenfeld and Emerson in this same volume) indicates that the factor which binds to ßP-F2 is avian Sp1 (22). Assuming that ßP-P2 is Sp1, how does that fact impact on the model? The only resultant change in our perception would be the perhaps surprising conclusion that a well known constitutive transcription factor (Sp1) is developmentally regulated for the timing of activation during erythrocyte maturation (Fig. 6), since it is obviously present at vastly reduced concentration (or activity) in primitive and immature definitive cells compared to mature definitive erythroid cells. The fact that the switch is not directly mediated by the erythroid-specific enhancer factor NF-E1 (since ßE-P4 is present in all three erythroid cell extracts; 23, 24) is perhaps not surprising, since NF-E1 expression may simply be required for establishment of the erythroid cell phenotype. As outlined in the model shown above then, the switch is theoretically accomplished by the collaborative interaction of tissue-specific and constitutive transcription factors; the regulation in the timing of expression of a constitutive trans-acting factor during red cell maturation is therefore likely to be one of the basic biochemical determinants of the switch.

ACKNOWLEDGMENTS. We thank our colleagues in the laboratory for many fruitful discussions, H. Beug (E.M.B.L.) for continued tutilege regarding the fine points of erythroblast cell culture, and B. Emerson (Salk Institute) for initially showing us how to prepare functional whole cell extracts. This work was supported by NIH grants HL24415 and GM28896 (J.D.E.), and by NIH NRSA postdoctoral fellowships to D.J.E. and J.L.G.

REFERENCES

1) Groudine, M., M. Peretz and H. Weintraub (1981) Molec. Cell Biol. 1:281-288.
2) Bruns, G. A. and V. M. Ingram (1972) Phil. Trans. Royal Soc. London (B) 266:225-305.

3) Choi, O.-R. and J. D. Engel (1986) Nature 323:731-734.
4) Hesse, J. E., J. M. Nickol, M. R. Lieber and G. Felsenfeld (1986) Proc. Nat. Acad. Sci. USA 83:4312-4316.
5) Behringer, R. R., R. E. Hammer, R. L. Brinster, R. D. Palmiter and T. M. Townes (1987) Proc. Nat. Acad. Sci. USA 84:7056-7060.
6) Kollias, G., J. Hurst, E. deBoer and F. Grosveld (1987) Nuc. Acids Res. 15:5739-5747.
7) Trudel, M. and F. Costantini (1987) Genes Devel. 1:954-961.
8) Grosveld, F., G. Blom van Assendelft, D. Greaves and G. Kollias (1987) Cell 51:975-985.
9) Forrester, W.C., S. Takegawa, T. Papayannopoulou, G. tamatoyannopoulos and M. Groudine (1987) Nuc. Acids Res. 15:10,159-10,177.
10) Choi, O.-R. B. and J. D. Engel (1988) Cell 55:17-26.
11) Nickol, J. M. and G. Felsenfeld (1988) Proc. Nat. Acad. Sci. USA 85:2548-2552.
12) Dolan, M., J. B. Dodgson and J. D. Engel (1983) J. Biol. Chem. 258:3983-3990.
13) Dodgson, J. B., S. J. Stadt, O.-R. Choi, M. Dolan, H. D. Fischer and J. D. Engel (1983) J. Biol. Chem. 263:12,685-12,692.
14) Emerson, B. M., C.D. Lewis and G. Felsenfeld (1985) Cell 41:21-30.
15) Beug, H., S. Palmeiri, C. Freudenstein, H. Zentgraf and T.Graf (1982) Cell 28:907-919.
16) Emerson, B. M., J. M. Nickol, P. D. Jackson and G. Felsenfeld (1987) Proc. Nat. Acad. Sci. USA 84:4786-4790.
17) Nakajima, N., M. Horikoshi and R.G. Roeder (1988) Molec. Cell Biol. 8:4028-4040.
18) Santoro, C., N. Mermod, P. C. Andrews and R. Tjian (1988) Nature 334:218-224.
19) Mitchell, P. J., C. Wang and R. Tjian (1987) Cell 50:847-861.
20) Imagawa, M., R. Chiu and M. Karin (1987) Cell 51:251-260.
21) Weintraub, H., H. Beug, M. Groudine and T. Graf (1982) Cell 28:931-940.
22) Lewis, C. D., S. P. Clark, G. Felsenfeld and H. Gould (1988) Genes and Devel. 2:863-873.
23) Wall, L., E. deBoer and F. Grosveld (1988) Genes and Devel. 2:1089-1100.
24) Evans, T., M. Reitman and G. Felsenfeld (1988) Proc. Nat. Acad. Sci. USA 85:5976-5980.

Hemoglobin Switching, Part A: Transcriptional Regulation, pages 105–116
© 1989 Alan R. Liss, Inc.

ACTIVATION MECHANISMS OF THE XENOPUS β GLOBIN GENE

Roger Patient, Mark Leonard, Alison Brewer, Tariq Enver, Angus Wilson and Maggie Walmsley. Department of Biophysics, Cell and Molecular Biology, University of London, King's College, 26-29 Drury Lane, London WC2B 5RL, UK.

INTRODUCTION

Gene activation involves the binding of non-histone proteins and changes in chromatin structure. The temporal relationship between these events is as yet unclear but the observed outcome is the loss of a nucleosome to yield a region hypersensitive to attack by nucleases, a so-called hypersensitive site (HS), within which a number of non-histone proteins are bound (1). In the case of the gene promoter, many of these proteins are transcription factors. A crucial requirement for a full understanding of the mechanisms of gene activation is to determine whether non-histone protein binding causes nucleosome loss or whether HS formation precedes protein binding.

Two candidate mechanisms for nucleosome removal are DNA replication and DNA supercoiling, either of which could be transient, requiring protein binding to prevent the nucleosome from returning. Here we present evidence implicating both of these processes in the activation of the Xenopus β globin gene. In addition we investigate the non-histone proteins, present in erythroid and non-erythroid cells, which bind to the Xenopus adult β globin gene promoter. By deletion analysis, we examine the sequences in the promoter responsible for HS formation.

COMPARATIVE FOOTPRINTING ON THE XENOPUS β GLOBIN GENE PROMOTER USING ERYTHROBLAST, ERYTHROCYTE AND NON-ERYTHROID NUCLEAR EXTRACTS

```
                                                          . . . . . . . . . . .
-474    AATTCCAGGCATATAAAAACACATTTATAAAAAACACTACATATACATATATATAT
        . . . . . . . . . . . . . . . . . . . . . . . . . . . . . .
-416    ATATATATATATATATATATATATATATATATATAAGTTTAAAAAGTGTGTTAATTTATA

-356    ATGTCTCTCTGGAAATAGAATTTCACACTTCAXTGTATACAAAATTATTAATATTTGTAA


-296    TATTTGATTATATTATGTTGTAGGGATAAAATGAATACCAGGCATATAAAAACACACTTT
                                                              └────────
                                     ┌──────────────────┐┌───
-236    AAAAAAAAAAAAAAATACATAGATACATAGATAATAATAATTTGTATTTATTTTTTCTTAA
        └──────────────────────────┘ └────────────────

                 HS
        ─┐     ┌→
        ─┘   └─┌─────────────────────────────────────┐┌───    ─┐ ┌──
-176    TATTCTAGCTCTGCTGTAATAAAAAAAAACATGCATCTAAAAGTGGTGCCAAATGGGAGGG
        ─┘              └──────────────────────┘└──────────┘└─         ┘
                                                                  OCTA

        ──────────────┐┌───────────────┐ ┌──────────────┐┌───
-116    TACAAATGGGCTGGGCAAATGTAACGTGTGCTTATCCTAGCCAATCAACAGGCAGAGTGG
        └──────────────┘└───────────┘└──────────┘└────────     ┌─┘└─
             OCTA           OCTA         GGATAAG     CCAAT       ^

        ──────────┐ ┌──────────────────────────────────────┐   +1
                                                                ↓↓    ↓
-56     AAAGGGGCAGTGCATCCTTACAGCTACATAAAGTCTGATGGATGGAGAATTAGAGCACTT
        ─────────────┘ └──────────────┘ └──────────────────────┘^↑↑  ↑↑

             SSE                    TATA


        ┌────────────────┐ ┌───────────────┐                          K.R.
+5      GTTCTTTTTGCAGAAGCTCAGAATAAACGCTCAACTTTGGCCATGGGTTTGACAGCACAT
        └────────────────┘ └───────────────┘       MetGlyLeuThrAlaHis
                                                       ┘                Ebs
                                                    └─
                                                     HS
```

Ebs = Erythroblasts ; K.R. = Kidney Cells ; HS = Nuclease

Hypersensitive Site ; ⌐‾⌐ = Reduced DNase I cleavage ;

↑ = Enhanced DNase I cleavage .

Fourteen days after the induction of anaemic shock with phenyl hydrazine, the peripheral circulation of adult Xenopus primarily contains transcriptionally active erythroblasts (2). A summary of the footprinting patterns observed with nuclear extracts from these cells is presented in Fig 1. In addition to protection of the TATA and CCAAT boxes, we observe three discrete footprints over a repeated sequence resembling the octamer motif. An octamer binding factor (OBF) has also been reported to bind to the human gamma globin promoter (6).

Of particular interest is the footprint immediately upstream of the CCAAT factor, where a good consensus AGGATAAG sequence is found. Thus Xenopus β, like every other globin gene so far examined, appears to have a binding site for the erythroid-specific protein Eryf1 (7) or NF-E (8). Neither mature erythrocytes nor non-erythroid cells appear to contain this protein (Fig 1 and data not shown), consistent with a role in gene activation. With extracts from the non-erythroid Xenopus cell line (KR, Fig 1), this sequence is protected by an extension of the CCAAT footprint in a manner reminiscent of the CCAAT displacement protein (CDP; 8,9).

Finally, it is also worth noting that the protected region immediately downstream of the CCAAT footprint contains some homology with the sequence in the chicken adult β globin promoter thought to be responsible for the selectivity of adult β globin gene expression in adult red blood cells (Fig 1; 10). This stage selector element (SSE)

Fig. 1. *Summary of comparative DNaseI footprinting on the Xenopus β globin promoter, using Xenopus erythroblast and kidney cell line extracts. Extracts were prepared by the method of Gorski et al (3), and footprinting (4,5) was performed on an end-labelled EcoRI-NcoI fragment from the promoter region. Approximately 100 µg of extract was incubated with 1 ng of labelled fragment, in the presence of 1-2 µg of cold poly (dI-dC) competitor. DNaseI digestion was usually performed with increasing amounts of enzyme. Regions of protection and enhanced cleavage are indicated and where a known consensus homology is apparent, this is marked. The boundaries of the chromatin DNaseI HS are also indicated.*

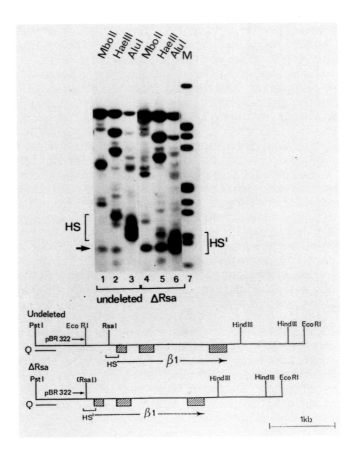

Fig. 2. *Restriction enzyme analysis of the hypersensitive site produced in the deleted and undeleted Xenopus β globin promoter. Nuclei from HeLa cells transfected with the SV40 Xenopus β globin hybrid plasmids, pJYMβ1Δ1pM or pJYMβ1Δ1pM ΔRsa (17) were digested with 3 units of MboII (lanes 1 and 4), 0.5 unit of HaeIII (lanes 2 and 5) or 0.5 unit of AluI (lanes 3 and 6), in 20 μl of buffer A (11) at 37°C for 30 min. Purified DNA was digested with PstI, and analysed by indirect end-labelling, using probe Q. The pBR322 MboII and HaeIII sites that become hypersensitive in ΔRsa (arrowed bands), are located 7 and 17 bp from the EcoRI site. Lane 7 is lambda DNA digested with EcoRI and*

is similarly located within the Xenopus and chicken promoters.

THE UPSTREAM BOUNDARY OF THE PROMOTER HYPERSENSITIVE SITE IS DETERMINED BY SEQUENCES MORE THAN 50 BP AWAY

We have previously reported that the Xenopus β globin promoter is active in transfected HeLa cells when linked to the SV40 enhancer and that transcriptional activation is accompanied by the formation in the chromatin of a nuclease hypersensitive site (HS) in this region (11,12). Studies with mammalian β globin genes in transfected HeLa cells have shown that the CACCC, CCAAT and TATA motifs are sufficient for transcriptional activity (13-16). As a first step towards defining the roles of the various promoter binding proteins in HS formation, we transfected a promoter mutant of the Xenopus β globin gene, deleted to -116. This mutant retains the TATA and CCAAT motifs, and two regions of CACCC homology centred around -105 and -89 (Fig 1). In agreement with the mammalian studies, this promoter was transcriptionally active in HeLa cells (17). In addition, the HS was formed and exhibited an unaltered upstream boundary at around -170, even though the Xenopus sequences in this region had been replaced by plasmid sequences. This is best illustrated using restriction enzymes with recognition sites in the plasmid DNA which becomes juxtaposed to the remaining Xenopus promoter sequences in the deletion mutant (Fig 2). These sites are not hypersensitive in the undeleted construct but become hypersensitive in the deleted templates. Thus, if the boundary of an HS reflects the resumption of nucleosomal structure, then sequences downstream of -116 direct nucleosome placement in the -170 region, that is from at least 50 bp away.

Footprinting studies on the Xenopus β globin promoter using HeLa cell extracts reveal DNase I cutting in the middle of the -105 CACCC homology (data not shown), suggesting that a CACCC factor does not bind there. A footprint is found over the weaker -89 CACCC homology but this region also contains sequences resembling the octamer consensus. Which of the remaining DNA sequences or their cognate proteins are required for HS formation may be determined by further mutagenesis studies.

HindIII, and 3' end-labelled. From Brewer et al (17), with permission.

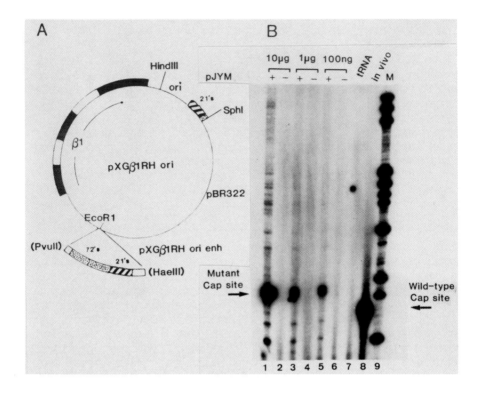

Fig. 3. Titration of transfected SV40-β globin plasmid in the presence (+) or absence (−) of 1 μg of the replication helper construct (pJYM). **Panel A:** Diagram of the SV40 origin and enhancer containing β globin construct used (pXGβ1RHorienh). The SV40 origin of replication (ori), 21 bp repeats (21's) and 72 bp repeats (72's) are marked. **Panel B:** Transcriptional analysis by primer extension of RNA from HeLa cells dextran transfected with the indicated amounts of pXGβ1RHorienh with or without pJYM. Total cytoplasmic RNA (350 ng) from Xenopus erythroblasts was used in the in vivo lane (lane 8). The pXGβ1RHorienh extension product differs from the in vivo product because of an in-phase 12 bp linker in the first exon of the β gene. M (lane 9) is 3' end-labelled HpaII-digested pAT153. From Enver et al (18), with permission.

β GLOBIN GENE ACTIVATION IN A NUMBER OF TRANSFECTED CELL LINES IS DEPENDENT ON DNA REPLICATION

In addition to the SV40 enhancer, transcription of the Xenopus β globin gene in HeLa cells requires linkage to a functional origin of replication. Fig 3 shows the transcriptional analysis of HeLa cell transfections in which a construct containing the Xenopus β globin gene linked to the SV40 enhancer and origin of replication was titrated over a 100-fold range in the presence or absence of a co-transfected SV40 T-antigen producing plasmid. T-antigen activates the SV40 origin of replication. Correctly initiated β globin transcripts were only detected, even at the highest level of transfected β globin gene, in co-transfections with the T-antigen producing plasmid: suggesting that replication is not merely acting by increasing the template copy number. This conclusion has been reinforced by co-transfection of replicating and non-replicating β globin constructs: only the replicating construct was transcriptionally active (18). Furthermore, to control for transactivation by T-antigen, the β globin gene was linked to a mutant origin of replication, which binds T-antigen but cannot initiate replication (19,20): β globin transcripts were not detected even in the presence of T-antigen (18).

Another potential explanation for the replication requirement of β globin transcription is the removal of bacterial methylation patterns from the template DNA. However, even when templates were replicated in transfected cells, rescued and retransfected, their transcription was dependent on replication in situ (18). Similar results have been obtained in the human B-cell line, Daudi, in the monkey kidney cell line, CV1 and in the constitutive T-antigen producing derivative of CV1, COS (data not shown). In addition, the replication requirement is independent of the position of the origin with respect to the gene (18), suggesting that the direction in which the replication fork passes through the gene is not critical. Thus, we conclude that replication is playing a mechanistic role in β globin gene activation, possibly by disrupting a stable preformed complex, such as a nucleosome, thereby permitting HS formation.

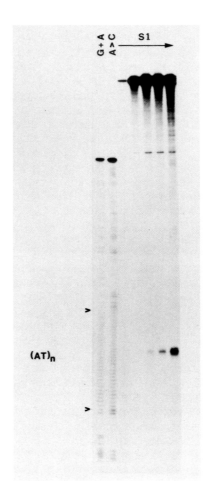

Fig. 4. High resolution mapping of S1 nuclease cleavage in an (AT)$_n$ tract in chromatin assembled in microinjected Xenopus oocytes. The plasmid was internally labelled at a site proximal to the (AT)$_n$ tract prior to injection. Injected oocytes were homogenised and digested with increasing amounts of S1 nuclease (5x10^{-4} to 5x10^{-2} units/ml) for 30 min at 37 $^{\circ}$C. Purified DNA was linearised at the site of internal labelling and recut to generate a uniquely end-labelled fragment containing the (AT)$_n$ tract.

THE SENSITIVITY TO NUCLEASE S1 OF AN ALTERNATING A-T TRACT
UPSTREAM OF THE β GLOBIN PROMOTER SUGGESTS THAT THE GENE IS
UNDER TORSIONAL STRESS

200 bp upstream of the β globin promoter HS is a 46 bp
stretch of alternating A-T residues (marked with dots in
Fig 1). We have previously shown that this region adopts
cruciform geometry at low levels of DNA supercoiling,
characterised by cleavage in the centre of the $(AT)_n$ tract
by the single strand specific nuclease, S1 (21,22). The
question of whether the DNA in chromatin is under torsional
stress has been controversial (23-27). Consequently, as an
independent test, we have probed the structure of the $(AT)_n$
sequence in microinjected Xenopus oocytes, where the
Xenopus β globin gene is assembled into chromatin and
transcribed very efficiently (28). We find that S1
nuclease cleaves specifically in the middle of the $(AT)_n$
tract (Fig 4), suggesting that the gene is indeed under
torsional stress. S1 nuclease requires Zn^{2+} ions and low
pH, conditions which could cause disruption of chromatin
structure, resulting in the production of free supercoiled
DNA circles, which would then be cleaved by S1. To test
whether this was the case, we performed micrococcal
nuclease digests on chromatin previously exposed to S1
buffer. The ladder of bands typical of nucleosomal arrays
was still detected suggesting that the chromatin had
survived intact (data not shown). In addition, when naked
DNA is exposed to prolonged S1 digestion, the pattern of
cleavage eventually spreads out from the centre of the
$(AT)_n$ tract (data not shown). We have not observed this
with prolonged digestion of chromatin, suggesting
protection of the surrounding sequences, possibly by
nucleosomes. We are currently testing the sensitivity of
the $(AT)_n$ tract to enzymes that will recognize the
supercoil-dependent structure in physiological buffer

---

*The products were fractionated on a 5% polyacrylamide gel
alongside the indicated sequencing reactions. The
boundaries of the $(AT)_n$ tract are marked. The data
presented here were obtained with the plasmid pXG540 (21),
which contains an $(AT)_{34}$ sequence from the Xenopus tadpole
αT1 gene. Similar results have been obtained with the
$(AT)_{23}$ sequence upstream of the Xenopus adult β globin
gene.*

conditions. However, from the experiments reported here, it would appear that actively transcribing β globin templates in microinjected oocyte nuclei are under torsional stress.

SUMMARY

Comparative protein binding studies have been performed on the Xenopus β globin gene promoter. Erythroblast nuclear extracts 'footprint' over the erythroid-specific consensus sequence, AGGATAAG, which is located immediately upstream of the CCAAT footprint. Non-erythroid cell extracts do not give rise to an AGGATAAG footprint but rather to an extended CCAAT footprint reminiscent of the CCAAT displacement protein (CDP). Erythroblast extracts also protect a sequence similar to the chicken stage selector element (SSE) immediately downstream of the CCAAT box footprint. In contrast to these discrete footprints observed using erythroblast extracts, Xenopus erythrocyte nuclear extracts give rise to more extensive promoter protection.

We have previously reported that this promoter is active in transfected HeLa cells when linked to the SV40 enhancer and that transcriptional activation is accompanied by the formation in the chromatin of a nuclease hypersensitive site (HS) in this region. As a first step towards defining the roles of the various promoter-binding proteins in transcriptional activation and HS formation, we transfected deletion mutants of the promoter into HeLa cells. Deletion of the sequences upstream of -116 had no effect on transcription or HS formation. Indeed the upstream boundary of the HS remained unchanged (at around -170) even though plasmid sequences had replaced Xenopus sequences. If the HS boundary reflects resumption of nucleosomal structure, then sequences downstream of -116 must be able to position a nucleosome from at least 50 bp away.

β globin gene activation in a number of transfected cell lines is absolutely dependent on DNA replication. The replication requirement is not a consequence of template copy number or methylation, nor is it dependent on the direction in which the replication fork passes through the gene. We conclude that replication facilitates active transcription complex formation by disrupting a stable association of the template with negative factors, which could include histones.

About 200 bp upstream of the Xenopus β globin gene promoter is a tract of alternating A and T residues which adopts cruciform geometry at low levels of supercoiling. Because of this sensitivity to torsional stress, we have probed the structure of the $(AT)_n$ sequence in microinjected Xenopus oocytes, where the Xenopus β globin gene is transcribed very efficiently. We find that S1 nuclease cleaves specifically in the middle of the $(AT)_n$ tract, suggesting that the gene is under torsional stress.

ACKNOWLEDGEMENTS

These studies were supported by the MRC, the Wellcome Trust and the SERC.

REFERENCES

1. Gross DS and Garrard WT (1988). Ann Rev Biochem 57:159–197.
2. Thomas N and Maclean N (1975). J Cell Sci 19:809–820.
3. Gorski K, Carneiro M and Schibler U (1986). Cell 47:767–776.
4. Galas DJ and Schmitz A (1978). Nucl Acids Res 5:3157–3170.
5. Emerson BM, Lewis CD and Felsenfeld G (1985). Cell 41:21–30.
6. Mantovani R, Malgaretti N, Giglioni B, Comi P, Cappellini N, Nicolis S and Ottolenghi S (1987). Nucl Acids Res 15: 9349–9364.
7. Evans T, Reitman M and Felsenfeld G (1988). Proc Natl Acad Sci USA 85:5976–5980.
8. Superti-Furga G, Barberis A, Schaffner G and Busslinger M (1988). EMBO J 7:3099–3107.
9. Barberis A, Superti-Furga G and Busslinger M (1987). Cell 50:347–359.
10. Choi O-R and Engel JD (1988). Cell 55:17–26.
11. Enver T, Brewer AC and Patient RK (1985). Nature 318:680–683.
12. Patient RK, Allan J, Brewer AC, Buckle R, Enver T and Walmsley M (1987). In: Developmental Control of Globin

Gene Expression, Prog Clin Biol Res 251:162–173.
13. Grosveld GC, van Ooyen A, de Boer E, Shewmaker CK and Flavell RA (1982). Nature 295:120–126.
14. Dierks P, van Ooyen A, Cochran M, Dobkin D, Reiser J and Weissmann C (1983). Cell 32:695–706.
15. Cochran MD and Weissmann C (1984). EMBO J 3:2453–2459.
16. Myers RM, Tilly K and Maniatis T (1986). Science 232:613–618.
17. Brewer AC, Enver T, Greaves DR, Allan J and Patient RK (1988). J Mol Biol 199:575–585.
18. Enver T, Brewer AC and Patient RK (1988). Mol Cell Biol 8:1301–1308.
19. Gluzman Y (1981). Cell 23:175–182.
20. Gluzman Y, Sambrook JF and Frisque RJ (1980). Proc Natl Acad Sci USA 77:3898–3902.
21. Greaves DR, Patient RK and Lilley DMJ (1985). J Mol Biol 185:461–478.
22. McClellan JA, Palacek E and Lilley DMJ (1986) Nucl Acids Res 14:9291–9309.
23. Cotten M, Bresnahan D, Thompson S, Sealy L and Chalkley R (1986). Nucl Acids Res 14:3671–3686.
24. Gottesfeld JM (1986). Nucl Acids Res 14:2075–2088.
25. Petryniak B and Lutter LC (1987) Cell 48:289–295.
26. Kmiec EB (1988). Cell 54:919.
27. Worcel A (1988). Cell 54:919.
28. Walmsley ME and Patient RK (1987). Development 101:815–827.

Hemoglobin Switching, Part A: Transcriptional Regulation, pages 117–127
© 1989 Alan R. Liss, Inc.

GENETIC AND BIOCHEMICAL ANALYSIS OF THE MOUSE β-MAJOR GLOBIN PROMOTER

Richard M. Myers, Alison Cowie, Laura Stuve, Grant Hartzog and Karin Gaensler

Departments of Physiology and Biochemistry & Biophysics, University of California at San Francisco, 513 Parnassus, San Francisco, CA 94143

INTRODUCTION

The mammalian globin genes provide a useful model system for studying transcription in eukaryotic cells. The expression of the genes is normally tightly restricted to just one cell type, the erythroid cell, so questions about the mechanisms of tissue-specificity can be addressed. Furthermore, because a controlled program of transcription of different globin genes occurs during development, the globin gene family offers a system to study switches in gene expression during embryonic and fetal development. Finally, as with most genes, it should be possible to use the promoters of globin genes to study the mechanisms by which the basic transcription machinery initiates RNA synthesis. Our efforts have been concentrated mostly on studying the mouse β-major globin promoter, with the aim of understanding the basis of the tissue-specificity of the gene, as well as answering some basic questions about mammalian transcription. Although we are not directly studying the developmental control of globin gene expression, it is conceivable that our studies of the adult beta-globin promoter will also provide some insight into the fetal-to-adult switch.

Most of our work has involved the use of Friend virus-transformed murine erythroleukemia (MEL; Friend et al., 1971) cells, which are a stable cultured pre-erythroid cell line that can be induced to differentiate into a cell type closely resembling erythroid cells by addition of a variety of chemical agents to the medium. After the addition of inducing agent to the cells, many changes that mimic red-cell differentiation occur, including a 20- to 50-fold increase in the initiation of transcription of the α- and β-globin genes. Our work with MEL cells has involved two experimental approaches. First, we have used DNA transfection of MEL cells to identify and characterize cis-acting DNA sequences within the mouse β-major globin promoter region that contribute to transcription of the gene. The second approach has been to identify, characterize and purify protein components present in nuclei of MEL cells that interact with these transcriptionally-important elements in the promoter. Our goal is to understand how these protein:DNA interactions, as well as potential protein:protein interactions between trans-acting components, contribute to the initiation of transcription of the gene. This paper describes our progress towards achieving this goal.

BACKGROUND RESULTS

Characterization of Cis-acting DNA Sequences

    In earlier studies with Drs. T. Maniatis and K. Tilly (Myers et al., 1986), we generated a large number of single base mutations within the mouse β-globin promoter region and studied the effects of these mutations on transcript levels in transiently-transfected HeLa cells. These studies identified three regions in the β-globin promoter that appear to play a role in transcription in these non-erythroid cells: (1) the "CACCC box" at -95 to -87 relative to the CAP site, (2) the CAAT box at -79 to -72, and (3) the TATA box at -31 to -26 (Figure 1). In addition, a minor decrease in the level of accumulated β-globin transcripts was observed with promoters carrying a single base change at the CAP site.

FIGURE 1: Key elements in the mouse β-major globin promoter region. The CACCC, CAAT and TATA boxes are indicated by boxes. A fourth element, which is an erythroid-specific <u>cis</u>-acting sequence that we have termed βDRE, is indicated by the directly repeating arrows. The numbers above the boxes refers to the nucleotide positions relative to the CAP site.

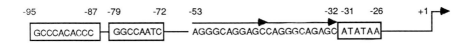

More recently, we determined the effects of many of these same single base mutations in the β-globin promoter on transcript levels in uninduced and induced MEL cells (Cowie and Myers, 1988). These experiments were initiated in the interest of examining the promoter in a cell type that more accurately represents the natural transcriptional background for globin genes. In order to study effects of many mutations on transcription of the β-globin genes in MEL cells, we developed transfection conditions that allow the uptake and expression of cloned globin genes in the cells in a transient manner. Unlike results that we obtained in long-term assays in which the transfected DNA is integrated into the chromosomal DNA in MEL cells, the transient assay is highly reproducible; therefore, it is better suited for examining the effects of many mutations on transcription levels. With this assay, we showed that the mouse β-globin promoter region driving a mouse metallothionein (MT) gene is expressed at low levels prior to induction of the cells with DMSO, and the levels increase 10- to 20-fold upon induction. Analysis of single base mutations in the promoter region gave results similar to those obtained in HeLa cells (Myers et al., 1986); three regions, the CACCC box, the CAAT box, and the TATA box were found to be important for transcript levels in both induced and uninduced MEL cells. One interesting difference was that two separate G to A transitions just upstream from the CAAT box, which showed 3- to 4-fold increases in transcription in HeLa cells, were neutral in MEL cells.

These studies also reported deletion experiments that indicated that the mouse β-globin promoter region from -106 to +26 carries all the 5'-flanking sequences needed to give maximal transcript levels in both induced and uninduced MEL cells (Cowie and Myers, 1988). β-globin/MT genes carrying 1200, 300 and 106 bp of 5'-flanking mouse β-globin sequences all resulted in the same levels of accumulated transcripts in both the transient and long-term assays. These results are at odds with reports implicating cis-acting sequences upstream from the human β-globin promoter in the -200 region (Antoniou et al., 1988). This discrepancy may be explained by genuine differences in the regulation of mouse and human β-globin genes; an argument against this explanation is that the -200 regions from the two species are similar. Other possible explanations for these results include differences in subclones of MEL cell lines and different transfection protocols. Regardless of the reasons for these differences, the behavior of the β-globin promoter region (as defined by our -106 construct) in the transient MEL assay suggests that it will be possible to study some of the many complex cis-acting elements in and around the gene separately.

RESULTS AND DISCUSSION

Identification of an Additional Cis-acting DNA Sequence in the Mouse β-globin Promoter Region

The transient expression assay developed for MEL cells described above suggested that a region of the mouse β-major globin promoter, which lies just upstream of the TATA box in the -55 to -32 region, plays a role in transcription of the gene; small effects on transcription could be seen when single base mutations in this region were examined in the assay. We found these results intriguing since the region contains a sequence of about 9 bp that is directly repeated in the promoter (Figure 2).

FIGURE 2: Nucleotide sequence and location of the βDRE, a directly repeated sequence in the mouse β-major globin promoter region. Double mutant promoters, containing two nucleotide substitutions within the repeat boxes, are shown below the sequence.

MUTANT	AGGGCAGGAGCCAGGGCAGAGC -53 -32
-47A/-35T	A T
-50A/-41C	A C
-46A/-40A	A A

This element appears, in directly repeated form, in the promoters of all mammalian adult β-globin promoters examined to date (Hardison, 1983). To analyze this region further, we constructed a set of "double" mutant promoters containing two separate single base substitutions, one in each of the repeated sequences (Figure 2). Cloned DNA templates carrying these mutant promoters driving the mouse MT gene were transfected into MEL cells and HeLa cells. The amounts of accumulated RNA corresponding to correctly initiated-transcripts of these genes were quantified by an S1 nuclease protection assay, using a cotransfected β-globin/neomycin gene fusion or a human α-globin gene as a standard. Double mutations in the two repeat boxes caused a decrease in the transcript levels of between 2- and 5-fold in induced MEL cells. Interestingly, these same mutations had no effect on transcript levels of the genes when transfected into HeLa cells, which are not of erythroid origin. These results suggest that this repeat region plays a role in transcription of the β-globin gene, and that it may be an erythroid-specific element. We have termed this cis-acting sequence the "βDRE" (β-globin Direct Repeat Element).

Trans-acting β-globin Transcription Components

To understand further the events leading to the initiation of transcription at the β-globin promoter, we have begun to study proteins present in the nuclei of MEL cells that interact with each of the cis-acting elements identified in the studies described above. The approach used is similar for each protein; MEL cells are grown in suspension, and extracts containing nuclear proteins are prepared from the cells either before or after inducing them to differentiate. The crude extracts are reacted with labeled DNA sequences carrying one or more of the DNA segments corresponding to the transcriptionally important elements identified in the transfection experiments described above. Interactions between components present in the extracts with the cis-acting DNA elements are observed by using the "mobility shift" assay (Fried and Crothers, 1981), and specificity is determined by various competition experiments. The interaction of each protein with its binding site is then further characterized by DNase footprinting experiments (Galas and Schmitz, 1978), as well as methylation interference experiments (Siebenlist and Gilbert, 1980); these studies are complemented by binding analysis to mutant promoter templates that are known to be defective in transcription in the transient MEL assay. The proteins are fractionated by conventional and binding site affinity chromatography, with the goal of purifying them to homogeneity. Finally, the binding proteins are used in reconstituted in vitro transcription experiments with protein fractions containing RNA polymerase to study the functional characteristics of the DNA binding activities. Below we discuss the identification of three proteins present in nuclei of MEL cells that interact with cis-acting sequences in the promoter, and our progress towards characterizing them. In addition, we describe the identification of a fourth protein present in MEL nuclei that interacts with a region of the mouse β-major globin gene within the second exon.

CAAT protein (βCP): The protein from MEL cell nuclei that we have characterized in most depth is one that interacts with the CAAT box region at -75 in the mouse β-major globin promoter (Cowie and Myers, 1989). This protein, which we have designated βCP, was first

identified in extracts from both induced and induced
MEL cells by mobility shift gels; specificity was
demonstrated by competition with other DNA sequences,
including additional CAAT boxes from other genes. We
found that the factor that binds efficiently to the β-
globin CAAT box also binds to a lesser degree to the α-
globin and Friend virus LTR CAAT boxes, but does not
bind to the CAAT boxes of the thymidine kinase or
histone H4 genes, nor to the NFI CAAT-like site (Jones
et al., 1987; Santoro et al., 1988). Promoters carrying
β-globin CAAT boxes with single base substitution
mutations that decrease transcription in the MEL
transfection experiments were also deficient in their
ability to bind to the βCP factor, which suggests that
the binding factor is involved in transcription of the
gene in MEL cells. The protein was fractionated to
near-homogeneity by conventional and affinity
chromatography, and a prominent silver-staining species
of molecular weight 38,500 was identified by
SDS/polyacrylamide gel electrophoresis. This species
was eluted from the gel, renatured and used in a band
shift gel assay to test for reconstituted binding
activity; this fraction was the only species in the gel
retaining binding activity upon renaturation.

The nearly-purified fraction of βCP was shown to
stimulate transcription specifically at the β-globin
CAAT box when added to transcription extracts that had
been depleted of βCP; this increase in transcription in
vitro was inhibited by addition of excess
oligonucleotides carrying the binding site, but not by
oligonucleotides composed of other sequences. These
results are consistent with the notion that βCP
contributes to transcription of the β-globin gene in MEL
cells by binding to the CAAT sequence. The lack of
binding to several other CAAT boxes, as well as the
distinct size of βCP suggests that it is different from
other CAT factors described in the literature (Jones et
al., 1987; Dorn et al., 1987; Landshulz et al., 1988;
Santoro et al., 1988; Chodosh et al., 1988); however,
our results do not indicate that βCP is erythroid-
specific. In fact, other non-erythroid cell types, such
as HeLa and L cells, contain a binding activity with
properties similar to βCP; whether these activities
correspond to the same protein remains to be determined.

CACCC protein: We have identified a protein present in MEL cell nuclear extracts that binds to the CACCC region of the promoter (at nt -95 to -87). This protein may be similar to a CACCC binding protein identified in human fetal erythroid cells (K562 cells; Mantovani et al., 1988) by Ottolenghi and coworkers. Experiments similar to those for βCP indicate that the binding of this protein to the CACCC region is specific. Furthermore, experiments with mutant promoters demonstrated a strong correlation of binding affinity with transcription from the promoter in the transient MEL assay. Footprinting and interference assays have been performed to identify the detailed features of the binding site. The protein has been purified to about 10% of homogeneity, and UV-crosslinking experiments suggest that a protein species of molecular weight 54,000 corresponds to the binding activity. Depletion and readdition experiments indicate that the binding activity depends on the presence of Zn^{+2} ions. In addition, preliminary experiments suggest that the protein is phosphorylated, and that phosphorylation is required for DNA binding. Experiments leading to complete purification of the protein are underway. Our goal is to use the purified protein in in vitro transcription assays to study its role in initiation of transcription.

Other proteins: Two other proteins have been identified in nuclear extracts of MEL cells, and will be discussed here. The first is an activity that binds to the βDRE, the direct repeat element in the promoter just upstream from the TATA box (at nt -53 to -32) that was identified in the MEL transient assay to be important functionally. Band shift gel assays were used to identify the activity, and competition experiments indicate that the binding is specific for the βDRE. Interestingly, this binding activity is approximately five-fold more abundant in induced MEL cell extracts than in uninduced extracts; this result is consistent with the βDRE being an erythroid-specific transcription element. Our aim is to characterize the binding of this factor to its DNA element by footprinting and interference assays, and to purify the protein and study its function in vitro.

Another protein was identified in extracts from MEL

cell nuclei that binds to a region of the mouse β-major globin gene within the second exon (between nt 359 and 466; Konkel et al., 1978). We discovered this binding activity fortuitously when we used a DNA fragment containing exon 2 as a negative control in promoter binding experiments. Because the activity appeared to be quite abundant, we decided to characterize it further. Analysis of several nuclear extracts indicated that the binding activity is present in both induced and uninduced MEL cells, but absent in two other cell types of non-erythroid origin (HeLa and L cells). Footprinting and interference experiments indicated that the binding is specific and covers the sequence 5' GGTGATAACTG 3', which is present at nucleotides +370 to +380 relative to the CAP site. We note that this sequence is similar to that shown to be important for enhancer function in the chicken β-globin gene region, and that the binding activity may be similar or the same as the chicken and/or human β-globin enhancer binding factors identified by several other groups (Emerson et al., 1987; Wall et al., 1988). We are cautious in interpreting these binding results, since we do not yet know whether the binding site is important for transcription of the mouse β-globin gene. We are testing this notion by mutagenesis and MEL transfection experiments similar to those described above.

REFERENCES

Antoniou M, deBoer E, Habets G, Grosveld F (1988). The human β-globin gene contains multiple regulatory regions: identification of one promoter and two downstream enhancers. EMBO J 7: 377-384.

Cowie A, Myers RM (1988). DNA sequences involved in transcriptional regulation of the mouse β-globin promoter in murine erythroleukemia cells. Mol Cell Biol 8: 3122-3128.

Cowie A, Myers RM (1989). Purification and characterization fo a CCAAT binding factor that stimulates transcription from the mouse β-globin promoter. Genes and Devel: submitted.

Dorn A, Bollekens J, Staub A, Benoist C, Mathis D

(1987). A multiplicity of CCAAT box-binding proteins. Cell 50: 863-872.

Emerson BM, Nickol JM, Jackson PD, Felsenfeld G (1987). Analysis of the tissue-specific enhancer at the 3' end of the chicken adult β-globin gene. Proc Natl Acad Sci USA 84: 4786-4790.

Fried M, Crothers DM (1981). Equilibria and kinetics of lac repressor-operator interactions by polyacrylamide gel electrophoresis. Nucl Acids Res 9: 6505-6525.

Friend C, Scher W, Killand JG, Sata T (1971). Hemoglobin synthesis in murine virus induced leukemic cells in vitro: stimulation of erythroid differentiation by dimethyl sulfoxide. Proc Natl Acad Sci USA 78: 378-382.

Galas D, Schmitz A (1978). DNase footprinting: a simple method for the detection of protein-DNA binding specificity. Nucl Acids Res 5: 3157-3170.

Hardison, RC (1983). The nucleotide sequence of the rabbit embryonic globin gene β4. J Biol Chem 258: 8739-8744.

Jones KA, Kadonaga JT, Rosenfeld PJ, Kelly TJ, Tjian R (1987). A cellular DNA-binding protein that activates eukaryotic transcription and DNA replication. Cell 48: 79-89.

Konkel DA, Tilghman SM, Leder P (1978). The sequence of the chromosomal mouse β-globin major gene: homologies in capping, splicing and poly(A) sites. Cell 15: 1125-1132.

Landshulz WH, Johnson PF, Adashi EY, Graves BJ, McKnight S (1988). Isolation of a recombinant copy of the gene encoding C/EBP. Genes and Devel 2: 786-800.

Mantovani R, Malgaretti N, Nicolis, S, Giglioni B, Comi P, Cappellini N, Tiziana Bertoro M, Caligaris-Cappio F, Ottolenghi S (1988). An erythroid specific nuclear factor binding to the proximal CACCC box of a β-globin gene promoter. Nucl Acids Res 16: 4299-4313.

Myers RM, Tilly K, Maniatis T (1986). Fine structure genetic analysis of a β-globin promoter. Science 232: 613-618.

Santoro C, Mermod N, Andrews PC, Tjian R (1988). A family of human CCAAT-box binding proteins active in transcription and DNA replication: cloning and expression of multiple cDNAs. Nature 334: 218-225.

Siebenlist U, Gilbert W (1980). Contacts between E. coli RNA polymerase and an early promoter of phage T7. Proc Natl Acad Sci USA 77: 122-126.

Wall L, deBoer E, Grosveld F (1988). The human β-globin gene 3' enhancer contains multiple binding sites for an erythroid-specific protein. Genes and Devel 2: 1089-1100.

Hemoglobin Switching, Part A: Transcriptional Regulation, pages 129–138

NUCLEAR PROTEINS WHICH BIND THE HUMAN γ-GLOBIN GENE

Deborah L. Gumucio, Todd A. Gray, Kirsten L. Rood, Kristina L. Blanchard, and Francis S. Collins

Departments of Internal Medicine (F.S.C. and D.L.G.), Human Genetics (F.S.C., T.A.G. and K.L.R.) and Howard Hughes Medical Institute (F.S.C. and K.L.B.), University of Michigan, Ann Arbor, Michigan 48109

INTRODUCTION

The molecular mechanisms responsible for the human fetal to adult hemoglobin switch have not yet been elucidated. Presumably, these mechanisms involve the interaction of trans-acting proteins with cis-acting sequences located near the γ-globin genes. Clues as to the location of these regulatory sequences might be gleaned from the study of naturally occurring mutations associated with abnormal expression of these genes. Such a model is provided by non-deletion hereditary persistence of fetal hemoglobin (HPFH), a condition characterized by elevated expression of the γ-globin gene(s) in adult life. By sequencing DNA from these individuals, point mutations have been identified in the promoter regions of the over-expressed genes. In fact, when cloned into expression vectors containing the chloramphenicol acetyl transferase (CAT) gene, and tested in K562 cells, the mutant promoters direct 3-8 fold over-expression of CAT relative to a wild type promoter (Collins et al., 1986).

To further establish the functional effect of the HPFH point mutations, we have used gel retardation and footprinting strategies to determine the effect of these mutations on the binding of nuclear proteins from expressing and non-expressing cell lines to γ-globin promoter sequences. Nuclear extracts were prepared according to Dignam et al. (1983). Figure 1 shows the origin of fragments and oligonucleotides used in these binding studies.

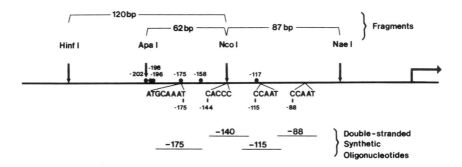

Figure 1. Promoter region of the human γ-globin genes from -299 to +35. Filled circles indicate sites of HPFH mutations.

CCAAT BOX BINDING PROTEINS

Figure 2A demonstrates that nuclear extracts from several cell types contain a protein (band 87A) which binds to the 87 bp NcoI/NaeI (-140 to -53) fragment which spans the duplicated CCAAT boxes. The protein was effectively competed by the addition of excess unlabelled synthetic oligonucleotides (30-mers) corresponding to the sequence of distal (γ-115) or proximal (γ-88) CCAAT boxes, but not by an oligonucleotide with a mutated CCAAT (→GGTAT) sequence (Fig. 2B). When γ-115 or γ-88 oligonucleotides were used as labelled probes, band 87A was detected with both, and was specifically competable (Fig. 2C). Thus, both distal and proximal CCAAT elements can bind to a ubiquitously distributed nuclear protein, designated γCAAT. To determine whether these and other CCAAT box sequences bind the factor with similar affinity, a careful titration of competition efficiency was performed (Fig. 2D). The comparative intensity of retarded bands was determined by laser densitometry and is plotted in Fig. 2E. The following heirarchy of binding affinity was determined: HPFH distal CCAAT (-117) > β-globin CCAAT > normal γ-globin distal CCAAT > γ-globin proximal CCAAT > HSV thymidine kinase CCAAT. Repeated experiments established that the -117 point mutation increased the affinity of the DNA-protein interaction by 2.27 + 0.26 (SEM, n=13) fold (p < 0.001).

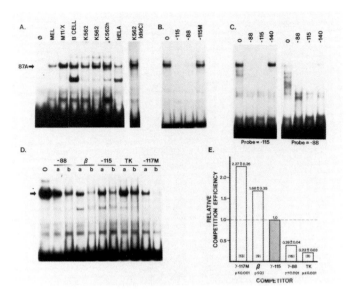

Figure 2. (A) Nuclear proteins which bind to the 87 bp
NcoI/NaeI fragment. Each extract (5-7 μg) was incubated
with 15,000 cpm (0.15 ng) of labelled 87 bp probe in the
presence of 500 ng of E. coli DNA. K562h - hemin induced
K562 cells (25 μM, 4 days). Last lane: poly(dI-dC) substituted for E. coli DNA. (B) Competition experiments. Lane
1: K562 extracts and the 87 bp NcoI/NaeI probe. Subsequent
lanes: 100 ng of indicated unlabelled oligonucleotide
preincubated with the K562 extract before probe addition.
(C) Gel retardation using synthetic oligonucleotide probes
derived from distal (γ-115, left) or proximal (γ-88, right)
CCAAT box regions. Unlabelled oligonucleotides (100 ng)
were added as specific competitor. -140 corresponds to an
unrelated oligonucleotide, from the CACCC region of the γ-
globin promoter. (D) Competition by various CCAAT containing sequences. K562 extract was used with the γ-115 probe.
Lane 1 contains no added specific competitor. Each subsequent pair of lanes contains 10 (a) and 50 (b) ng of the
oligonucleotide indicated. (E) Quantitative competition
results. Multiple gels such as the one shown in Fig. 2D
were scanned with a laser densitometer. All lanes were
compared to the density of bands obtained using the γ-115
(distal) competitor (set at 1.0). Means and standard
errors (S.E.M.) are given above each bar, and the number of
separate determinations appears in parentheses.

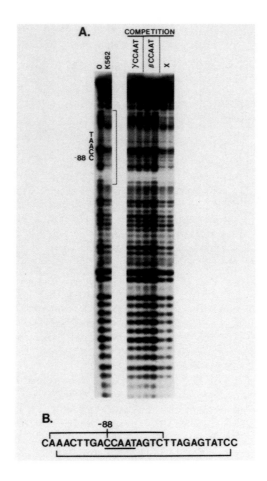

Figure 3. Direct DNaseI footprinting of a CCAAT box binding factor. (A) 0 = no protein added. K562 = 30 μg of crude nuclear extract incubated with 20,000 cpm (0.2 ng) of the end labelled 87 bp NcoI/NaeI probe and 2 μg of poly(dI-dC). The next 6 lanes show the results of competition studies. Twenty-five or 50 ng (in paired lanes) of unlabelled specific competitor DNA was added to the binding reaction: γCCAAT = γ-88; βCCAAT = sequence of the human β-globin gene; X = unrelated oligonucleotide (γ-140). (B) Summary of DNaseI protection results. Brackets indicate footprints produced by crude K562 extracts on sense (above the sequence) and antisense (below) strands.

The binding preference of γCAAT for globin CCAAT sequences over thymidine kinase CCAAT (Fig. 2E) is shared by CP1, one of three CCAAT box binding proteins found in HeLa extracts (Chodish et al., 1988). Thus γCAAT may be closely related or identical to CP1.

Using the K562 extract, a DNaseI footprint was detected over the proximal CCAAT box (Fig. 3A). The footprint was specifically competable by both the -115 and -88 CCAAT oligonucleotides, but not by a non-specific oligonucleotide. Although gel retardation experiments (Fig. 2D, E) showed that the affinity of γCAAT binding to the distal CCAAT box was approximately 2 fold higher than to the proximal CCAAT sequences, a footprint was not consistently detected over the distal CCAAT box. It is possible that this binding interaction is characterized by a high dissociation constant which is stabilized by gel retardation assays by the caging effect of the polyacrylamide.

OCTAMER BINDING PROTEINS

Gel retardation experiments using the 62 bp NcoI/ApaI (-202 to -140) fragment revealed ubiquitous as well as cell-type specific binding interactions (Fig. 4A). Band 62A was seen in all extracts tested; band 62B was detected strongly in B cells and weakly in K562 and HEL cells; bands 62C and 62D were restricted to erythroid extracts (MEL, M11/X, HEL and K562).

Two HPFH mutations have been described in the region spanned by this 62 bp probe, -158 (C → T) and -175 (T → C). When a fragment containing the -158 HPFH mutation was used as a probe, the results were identical to those described for the wild type probe. However, when a 62 bp fragment containing the -175 mutation was used, the binding of proteins in bands 62A and 62B was dramatically reduced (Fig. 4B). This mutation consists of a transition (T → C) in the eighth base of the octamer sequence, ATGCAAAT. This sequence has been shown to be a binding site for a transcriptional activator protein in immunoglobulin, histone, and a variety of other genes. Two forms of octamer binding protein (OBP) have been described (Landolfi et al., 1986; Staudt et al., 1986). Band 62A in Fig. 4A comigrates with the "ubiquitous" form of OBP while band 62B comigrates with the B-cell specific form. To further confirm that bands 62A and 62B correspond to these previously described octamer binding proteins, a ^{32}P-labelled oligonucleotide probe containing the V_H octamer sequence was incubated with the K562 extract. Fig. 4C shows that this probe formed complexes of identical mobility to bands 62A and 62B.

The binding of all four complexes to the wild type 62 bp probe was effectively competed by a synthetic 35-mer spanning the γ-globin octamer sequence. A 35-mer carrying the -175 mutation competed only for bands 62C and 62D. In contrast, the synthetic V_H octamer sequence competed for bands 62A and 62B, but not 62C and 62D. An unrelated 30-mer did not compete effectively for any of the K562 binding activities (data not shown).

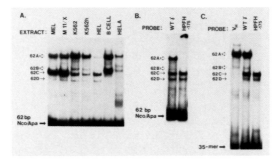

Figure 4.   Binding of Nuclear proteins to the octamer region.   (A) Various extracts (5-7 μg) were incubated with the 62 bp ApaI/NcoI fragment.   (B) Effect of the -175 HPFH mutation on protein binding.   Lane 1: binding of the K562 extract to the normal 62 bp fragment (γWT).   Lane 2: reduction in binding of bands 62A and 62B seen with the use of the -175 mutant 62 bp probe.   (C) Comigration of bands 62A and 62B with octamer binding proteins.   Oligonucleotide probes used were: V_H = octamer sequence from the immunoglobulin variable region heavy chain promoter; WTγ = 35-mer spanning the normal γ-globin octamer region; HPFH -175 = γ-globin 35-mer containing the -175 HPFH mutation.

The copper-orthophenanthroline (Cu/OP) cleavage reaction (Kuwabara & Sigman, 1987) was used to determine the sequences protected by each of these four binding proteins in K562 cells.   Fig. 5B shows that the B cell proteins in bands 62A and 62B produced identical footprints, protecting

16-18 bp over the octamer sequence of the 62 bp probe.
K562 band 62A also produced an identical footprint, as did
band 62B (not shown). In contrast, the footprint produced
by band 62C overlapped with and extended further 5' than
the octamer binding activity (Fig. 5C). Band 62D produced
a footprint identical to band 62C (data not shown). Direct
DNase I footprints were also obtained using the 120 bp
NcoI/HinfI fragment, Klenow labelled at the NcoI site (Fig.
5D). K562 extracts protected the same region as bands 62C
and 62D, while B cell extracts produced a footprint over
the octamer region.

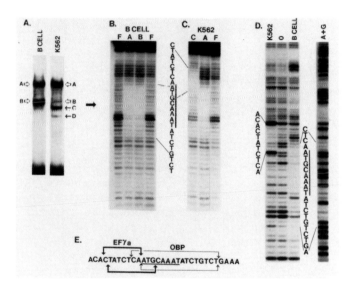

Figure 5. Footprinting of octamer and EFγa binding
proteins. (A) Gel retardation assay of the normal 62 bp
probe incubated with B-cell or K562 extracts. (B) Cu/OP
footprints of B-cell bands 62A and 62B. (C) Cu/OP
footprints of K562 proteins 62A and 62C. The extent of
protection of the noncoding strand is summarized at left.
(D) Direct DNaseI footprinting of the 120 bp NcoI/HinfI
fragment by K562 and B-cell extracts (40 μg). The center
lane contains probe digested in the absence of nuclear
extracts. The right lane shows the A + G sequencing
reaction of this region. (E) Summary of overlapping
binding of OBP and EFγa.

Fig. 5E summarizes Cu/OP footprinting studies done on both strands. We have designated the protein which binds the octamer γOBP. The protein which overlaps with the octamer binding site has been named EFγa (Erythroid Factor γa).

OTHER BINDING PROTEINS

The CACCC sequence, located at -144 of the γ-globin promoter, is highly conserved in all β-like globin genes. The functional importance of this element is demonstrated by the fact that point mutations within the human β-globin CACCC sequence result in β-thalassemia (Orkin et al., 1982). Other mutations in this element destroy its ability to activate transcription in transient expression assays (Myers et al., 1986). We have detected ubiquitously distributed proteins which bind the γ-globin CACCC sequence (Gumucio et al., 1988). Recent data from our laboratory indicate that at least one of these proteins is the Sp-1 protein first described by Dynan and Tjian (1983). Mutations in the CACCC element which decrease expression also severely reduce the binding of these proteins. Thus, the CACCC binding proteins may represent important factors for β-like globin gene expression.

Finally, we have detected additional proteins which bind to the -200 region of the γ-globin promoter. A cluster of HPFH mutations have been identified in this region at -202, -198, and -196. The fact that these mutations all produce the same phenotype suggests that this region may represent a binding site for a repressor protein. Indeed, the binding of at least one of the proteins we are studying is reduced by the -202 mutation (data not shown).

DISCUSSION

We have detected several proteins in human erythroleukemia cells (K562 and HEL) which bind to the region spanning -210 to -50 of the γ-globin gene promoter. The binding of three of these was modified by HPFH-associated sequence changes. In two cases, -175 (T → C), and -202 (C → G), the point mutation interfered with protein binding, suggesting that the sequences define repressor binding sites. In the third case, the -117 mutation increased the affinity for γCAAT, a factor which binds both γ-globin CCAAT boxes. In each case, the altered pattern of binding of these nuclear proteins could contribute to increased expression of the γ-globin gene in HPFH.

Of interest to the overall picture of γ-globin gene expression in the finding that two proteins, EFγa and a · ubiquitously distributed OBP bind to overlapping sequences within the promoter. The fact that the -175 HPFH mutation greatly reduces the affinity of γOBP binding suggests that this protein, which has been well described as a transcriptional activator, acts as a repressor in this context. It is tempting to speculate that differences in the availability of EFγa and γOBP or regulated modifications of either of these proteins could be important components of the switching mechanism.

Recent data (not shown) indicate that the erythroid specific EFγa protein is probably identical to the Eryf-1 factor detected in chicken erythrocytes (Evans et al., 1988) and the NF-E1 factor which binds the human β-globin gene (Wall et al., 1988). The tissue-specific distribution of this factor, the appearance of its binding site in several globin genes (Evans et al., 1988) as well as its apparent evolutionary conservation suggest that this protein may be an integral part of the transcriptional apparatus necessary for erythroid-specific globin expression.

REFERENCES

Chodosh LA, Baldwin AS, Carthew RW, Sharp PA (1988). Human CCAAT-binding proteins have heterologous subunits. Cell 8:11-24.

Collins FS, Bodine DM, Lockwood WK, Cole JL, Mickley L, Ley T (1986). Expression analysis of the human fetal globin gene promoter in hereditary persistence of fetal hemoglobin. Clin Res 34:454a.

Dignam JD, Lebowitz RM, Roeder RG (1983). Accurate transcription initiation by RNA polymerase II in a soluble extract from isolated mammalian nuclei. Nucleic Acids Res 11:1475-1489.

Dynan WS, Tjian R (1988). Isolation of transcription factors that discriminate between different promoters recognized by RNA polymerase II. Cell 32:669-680.

Evans T, Reitman M, Felsenfeld G (1988). An erythocyte-specific DNA-binding factor recognizes a regulatory sequence common to all chicken globin genes. Proc Natl Acad Sci USA 85:5976-5980.

Gumucio DL, Rood KL, Gray TA, Riordan MF, Sartor CI, Collins FS (1988). Nuclear proteins which bind the human γ-globin gene promoter: Alterations in binding produced by HPFH-associated point mutations. Molec Cell Biol (in press).

Kuwabara MD, Sigman DS (1987). Footprinting DNA-protein complexes in situ following gel retardation assays using 1, 10-phenanthroline-copper ion: <u>Escherichia</u> <u>coli</u> RNA polymerase-<u>lac</u> promoter complexes. Biochemistry 26:7234-7238.

Landolfi NF, Capra JD, Tucker PW (1986). Interaction of cell-type-specific nuclear proteins with immunoglobin V_H promoter region sequences. Nature 323:548-549.

Myers RM, Tilly K, Maniatis T (1986). Fine structure genetic analysis of a β-globin promoter. Science 232:613-618.

O'Hare P, Goding CR (1988). Herpes simplex virus regulatory elements and the immunglobulin octamer domain bind a common factor and are both targets for virion transactivation. Cell 52:435-445.

Orkin SH, Kazazian Jr. HH, Antonarakis SE, Goff SC, Boehm CD, Sexton JP, Waber PG, Giardina PJV (1982). Linkage of β-thalassemia mutations and β-globin gene polymorphisms with DNA polymorphisms in human β-globin gene cluster. Nature 296:627-631.

Preston CM, Frame MC, Campbell MEM (1988). A complex formed between cell components and an HSV structural polypeptide binds to a viral immediate early gene regulatory DNA sequence. Cell 52:425-434.

Staudt LM, Singh H, Sen R, Wirth T, Sharp PA, Baltimore D (1986). A lymphoid-specific protein binding to the octamer motif of immunoglobulin genes. Nature 323:640-643.

Wall L, deBoer E, Grosveld F (1988). The human β-globin gene 3' enhancer contains multiple binding sites for an erythroid specific protein. Genes and Devel 2:1089-1100.

Hemoglobin Switching, Part A: Transcriptional Regulation, pages 139–148
© 1989 Alan R. Liss, Inc.

Sites I and II Upstream of the $^A\gamma$ globin gene bind
nuclear factors and affect gene expression

Joyce A. Lloyd, Richard F. Lee, Anil G. Menon,
and Jerry B Lingrel
Department of Molecular Genetics, Biochemistry
and Microbiology, University of Cincinnati
231 Bethesda Avenue
Cincinnati, Ohio 45267-0524

INTRODUCTION

The globin genes in vertebrates are regulated in a
tissue and developmental specific manner. The human $^A\gamma$
globin gene is normally expressed in the fetal yolk sac,
though a low level of $^A\gamma$ globin is synthesized during
adult life. In individuals with hereditary persistence
of fetal hemoglobin (HPFH), however, $^A\gamma$ globin levels are
higher than normal in adults. The HPFH condition is
asymptomatic. It has been correlated with certain point
mutations in the region upstream of the $^A\gamma$ and $^G\gamma$ globin
genes, for example, at positions -117, -175, -196, and -
202 from the transcription start site (Collins et al.,
1985; Gelinas et al., 1985; Gelinas et al., 1986; Month
et al., 1986; Schwartz et al., 1987). Since fetal
hemoglobin is functional in adults which make no adult
hemoglobin, study of the molecular mechanisms by which
the HPFH mutations allow adult expression of the γ globin
gene is medically relevant.

In order to elucidate the molecular mechanisms of γ
globin regulation, we determined whether nuclear factors
from K562 erythroleukemic cells, which express the gene,
specifically bind to the $^A\gamma$ globin gene. The region
encompassing the HPFH mutations and other regions of the
gene were tested. Multiple DNA fragments spanning and
flanking the $^A\gamma$ globin gene were incubated with K562
nuclear extract and electrophoresed in band retardation
gel-binding assays. The entire $^A\gamma$ globin gene binding

map, as well as a detailed discussion of two particular binding regions upstream of the $^A\gamma$ globin gene follows. Footprints of bases involved in protein binding in these two regions have been obtained. The importance of the footprinted sequences for gene expression was determined in transient expression assays. To accomplish this, $^A\gamma$ globin - chloramphenicol acetyl transferase (CAT) fusion genes were constructed containing mutations of the bases which interact with protein in the footprint assay. These mutagenized constructs were transfected into K562 and other cells, and their levels of expression were compared to wild-type.

RESULTS AND DISCUSSION

Figure 1 shows the results obtained from mapping the entire $^A\gamma$ globin gene and flanking sequences for regions which specifically bind protein in band-retardation assays. A series of restriction endonuclease fragments spanning the gene were individually end-labeled at either the 5' end using γ-^{32}PdATP or at the 3' end using Klenow fragment and α-^{32}PdNTP. The end-labeled fragments were incubated with 3 µl of K562 nuclear extract, prepared as described by Dignam et al. (1983), in the presence of poly dI/dC in a reaction volume of 20 ul. Each fragment tested is indicated as a line below the gene map. The numbers above the fragments indicate length in base pairs, and the relative positions of the fragments with respect to the $^A\gamma$ globin gene introns and exons (boxes) are shown. Broad lines indicate fragments which bind K562 nuclear proteins with sequence specificity, that is, unlabeled DNA including the fragment in question can compete for protein binding better than the same quantity of DNA with a random sequence. In the first three lanes of Figure 2, a band retardation assay of the 182 bp

---

Figure 1 - K562 nuclear factor binding map of the human $^A\gamma$ globin gene. A schematic diagram of the $^A\gamma$ globin gene including the positions of exons and some of the promoter sequences is shown. Each fragment was end-labeled and tested in a gel retardation assay for sequence-specific DNA binding with factors from K562 nuclear extracts. The sizes in bp of the fragments are indicated. Bold lines indicate fragments which bind protein specifically.

Human Aᵧglobin

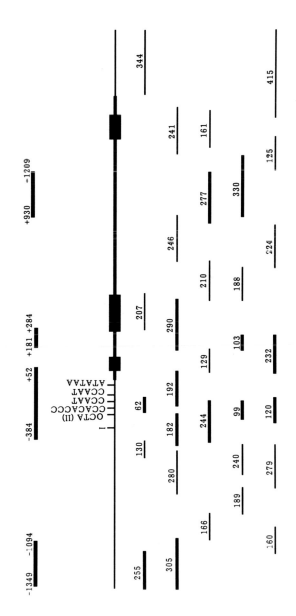

fragment spanning sites I and II (octa) upstream of the ᴬγ globin gene is shown, as an example. The fragment was incubated with varying amounts of poly dI/dC non-specific competitor, as indicated. The DNA which is not protein bound runs faster on the gel than that which is bound. In this case two DNA-protein complexes (B1 and B2) are evident in the gel retardation of the 182 bp fragment incubated with K562 nuclear extract.

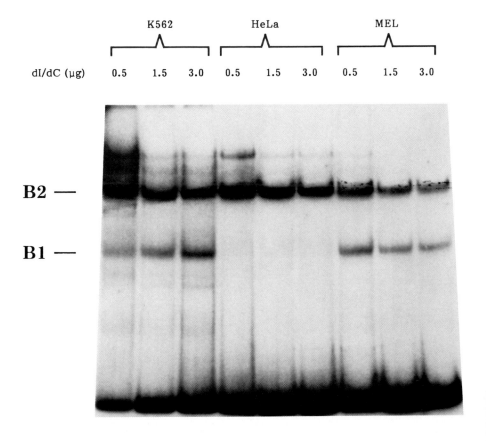

Figure 2 - Binding of the 182bp fragment with K562, HeLa and MEL extracts. The complexes formed by DNA-protein interactions are designated B1 and B2. Labeled DNA which is not protein-bound runs with a faster mobility than B1 or B2. Binding reactions included the indicated amounts of poly dI/dC.

In summary, the $^A\gamma$ globin gene binding map in Figure
1 reveals four broad regions of protein interaction. Two
are 5' of the gene, between -1349 and -1094, and -384 to
+52 from the transcription start site.   Other binding
regions are located in the first and second introns, +184
to +284 and +930 to +1209.   A fifth region which binds
protein is not shown on this map; it is located in the
750 bp fragment 3' of the $^A\gamma$ globin gene which has been
shown to have enhancer activity (Bodine and Ley, 1987).

These regions in and around the $^A\gamma$ globin gene are
capable of specifically binding nuclear factors from K562
nuclear extracts, and each may play a role in regulation
of expression.   The experiments do not reveal how many
binding sites are located within a fragment, i.e. there
may be two (or more) factors binding to the $^A\gamma$ globin
gene between -1349 and -1094, each at a different
location.   By footprinting the DNA-protein interactions,
it is possible to locate the bases within these fragments
which might be important for function.

We wished to study in greater detail the binding
region just 5' of the $^A\gamma$ globin gene.   It contains the
bases for which point mutations involved in HPFH occur,
and so may represent protein interactions responsible for
developmental control of $^A\gamma$ globin gene expression.   The
182 bp fragment used in the band retardation assay in
Figure 2 spans the region containing the nucleotides at -
175, -196 and -202, which are correlated with HPFH.   A
complex (B2) with similar mobility is evident in binding
assays with extracts from HeLa, MEL (mouse
erythroleukemic) or K562 cells.   This suggests that the
B2 complex is probably not erythroid-specific and
probably represents the binding of an evolutionarily
conserved protein.   In contrast to B2, the B1 binding
complex did not prove to be sequence specific by the
criteria previously mentioned, even though it does appear
only with the erythroid (K562 and MEL) extracts, and not
with HeLa.   The B1 complex was not further investigated.

Figure 3 depicts a schematic of the results of two
methods of footprinting assays performed with a fragment
spanning the region containing -202, -196 and -175.   For
both methods, preparative quantity binding assays were
performed and the free and protein-bound DNA were treated
with a cleaving agent, and subsequently run on sequencing

gels.   Both strands of the sequence are shown, and the
bases which footprint on each strand are in brackets.
Solid brackets indicate footprints obtained with the OP-
Cu (copper -1, 10 phenanthroline) chemical cleavage
method (Kuwabara and Sigman, 1987), dotted brackets
represent DNase I footprinting results. The arrows above
the coding strand of sequence represent positions of

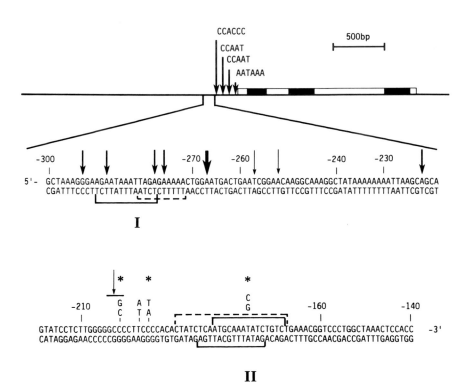

Figure 3 - Footprinting results for sites I and II.
DNase I footprints are stippled lines and solid lines are
OP-Cu footprints. Some of the characterized HPFH point
mutations are shown above the normal sequence
(asterisks). The arrows point to bases in the coding
strand which are more sensitive to OP-Cu cleavage in
protein-bound than in free DNA. Site I extends from -267
to -293 and site II from -168 to -185.

increased sensitivity to OP-Cu cleavage in protein-bound vs. free DNA. Increased boldness of the arrow indicates higher relative sensitivity. A cluster of sensitive sites between positions -267 and -293 coincides with the DNase I and Op-Cu footprints on the opposite strand. This binding region is designated site I. A second binding region is located between -168 and -185, as determined by OP-Cu footprinting on both strands. So, there are two DNA footprints near the positions of the HPFH mutations. Site I is upstream from the -202 position; site II is downstream of -202 and -196, but includes -175. Interestingly, site II also includes the octamer sequence (ATGCAAAT) found to be important in expression of the immunoglobulin and histone H2b genes (Sen and Baltimore, 1986; Sive and Roeder, 1986). We do not know, however, if the protein which binds site II is related to the H2b or immunoglobulin octamer binding factors.

Individual DNA fragments containing either site I or site II each show a pattern of gel retardation in binding assays. A clustered-base substitution made by in vitro mutagenesis of site I, or a -175 point mutation made in site II, decreased protein factor binding in band retardation assays (data not shown). This confirms that the bases which footprint are responsible for the observed gel shift.

We wished to determine the functional significance of the binding of K562 nuclear factors at sites I and II. The promoter region 5' of the $^A\gamma$ globin gene (-1350 to -32), with or without the mutations at site I or II, was fused to a CAT reporter gene. This fusion gene was electroporated into K562, HEL (another human erythroleukemic line), HeLa and 293 (adenovirus transformed human kidney) cells in transient expression assays. CAT assay results from a HEL cell transfection are shown in Figure 4. A clustered-base substitution in site I results in greatly decreased CAT expression, at least 20X, with respect to wild-type. This may indicate that the factor binding at site I normally has a positive effect on transcription. This decreased expression of site I mutants is apparent in all the cell types transfected, both erythroid and non-erythroid. The construct with the point mutation at -175 shows CAT activity 2-3X greater than wild-type. This correlates

with the condition of the naturally occurring HPFH. The
increase was seen only in erythroid cells, and not in 293
or HeLa cells. The simplest explanation for this
increase in expression would be that a negative regulator
can no longer bind due to the -175 (T→C) point
mutation. However, it is also possible that the point
mutation facilitates better binding of a positive factor,
or some combination of these two explanations.

HEL CELL
CAT ASSAY

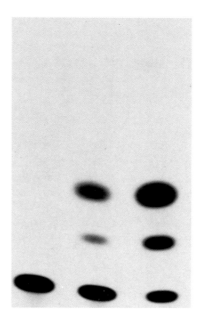

Construct
(γAdCAT)   SI   w.t.   -175 (T→C)

Figure 4 - A representative CAT assay. The
indicated constructs were electroporated into HEL cells
and CAT assays were performed after 48 hours. SI
represents a clustered-base substitution from -271 to -
280 in site I. The constructs include the promoter
region 5' of the Aγ globin gene fused to the adenovirus
major late promoter and in turn to the CAT gene. The
transciption start site is in the Ad MLP.

SUMMARY

Five broad regions in and near the $^A\gamma$ globin gene specifically bind proteins in K562 nuclear extracts. Two are located 5' of the gene, one between -1349 and -1094, and the other between -384 and +52. Each of the two introns has a binding region, +184 to +284 and +930 to +1209. The fifth binding region is within the $^A\gamma$ globin gene enchancer, a 750 bp Hind III fragment located 3' to the gene. The nuclear factors which bind to any or all of these regions may be important for control of regulation of the $^A\gamma$ globin gene.

Since HPFH point mutations occur at -175, -196 and -202, we investigated K562 nuclear factor binding to this region more closely. Footprints were obtained for two binding sites, one from -168 to -185 (site II) and the other from -267 to -293 (site I). Site II contains an octamer sequence (ATGCAAAT) important for protein binding and expression of the histone H2b and immunoglobulin genes. The HPFH mutation at -175 changes the T in the octamer sequence to a C. Site I is upstream of the bases affected by HPFH mutations.

A $^A\gamma$ globin gene promoter - CAT reporter fusion gene was constructed with a clustered - base substitution of site I or a T→C point mutation at -175. The mutation in site I decreases CAT expression 20X compared to wild-type in erythroid and non-erythroid cells. Site I probably binds a positive regulator. The T→C change at -175 increases expression 2-3X over wild-type in erythroid cells, but not in non-erythroid cells. This increase correlates with the effect of the naturally occurring HPFH, and may result from decreased binding of a negative effector, and/or increased positive factor binding.

REFERENCES

Bodine DM, Ley TJ (1987). An enhancer element lies 3' to the human $^A\gamma$ globin gene. EMBO J., 6: 2997-3004.

Collins FS, Metherall JE, Yamakawa M, Pan J, Weissman SM, Forget BG (1985). A point mutation in the $^A\gamma$-globin gene promoter in Greek hereditary persistence of fetal hemoglobin. Nature, 313: 325-326.

Dignam JD, Lebowitz RM, Roeder RG (1983). Accurate transcription intiation by RNA polymerase II in a soluble extract from isolated mammalian nuclei. Nucleic Acids Res., 11: 1475-1489.

Gelinas R, Bender M, Lotshaw C, Waber P, Kazazian, Jr., Stamatoyannopoulos G (1986). Chinese $A\gamma$ fetal hemoglobin: C to T substitution at position -196 of the $A\gamma$ gene promoter. Blood, 67: 1777-1779.

Gelinas R, Endlich B, Pfeiffer C, Yagi M, Stamatoyannopoulos G (1985). G to A substitution in the distal CCAAT box of the $A\gamma$-globin gene in Greek hereditary persistence of fetal hemoglobin. Nature, 313: 323-325.

Kuwabara MD, Sigman DS (1987). Footprinting DNA-protein complexes in situ following gel retardation assays using 1,10-phenanthroline-copper ion: Escherichia coli RNA polymerase-lac promoter complexes. Biochemistry, 26: 7234-7238.

Month S, Delgrosso K, Orchowski P, Rappaport E, Malladi R, Schwartz E, Surrey S (1986). Analysis of the region 5' to the $G\gamma$ gene in a patient with $G\gamma$-β+-HPFH/βs. Blood, 68: 76a.

Schwartz E, Month S, Delgrosso K, Rappaprot E, Ochowski P, Malladi P, Surrey S (1987). DNA sequence changes in the 5'-flanking region of $G\gamma$-globin genes in a black with βs and a non-deletional form of $G\gamma$-β+ HPFH. In Stamatoyannopoulous and Nienhuis (eds.), Developmental Control of Globin Gene Expression. Alan, R., Liss, Inc., New York, pp. 363-371.

Sen R, Baltimore B (1986). Multiple nuclear factors interact with the immunoglobulin enhancer sequences. Cell 46: 705-716.

Sive H, Roeder RG (1986). Interaction of a common factor with conserved promoter and enhancer sequences in histone H2b, imunoglobulin, and U2 small nuclear (sn)RNA genes. Proc. Natl. Acad. Sci. USA, 83: 6382-6386.

Hemoglobin Switching, Part A: Transcriptional Regulation, pages 149–162
© 1989 Alan R. Liss, Inc.

TRANSCRIPTIONAL REGULATION OF HUMAN GAMMA GLOBIN GENE EXPRESSION

Kevin T. McDonagh, Henry Lin, David M. Bodine,
Christopher Lowrey, Mary Purucker, Timothy Ley
and Arthur W. Nienhuis, Clinical Hematology Branch
NIH, Bethesda, MD and Department of Medicine,
Washington University School of Medicine,
St. Louis, MO (TL)

INTRODUCTION

The globin genes are a model system for studying strict tissue specific and developmentally specific regulation of gene expression. The switch that occurs in humans from γ globin to β globin during the perinatal period is of both basic scientific and clinical interest. An understanding of the molecular mechanisms responsible for this switch may aid in the development of genetic strategies for the treatment of β globin disorders such as thalassemia and sickle cell disease.

A variety of experimental models have identified two distinct types of cis-acting regulatory elements, promoters and enhancers, which are in part responsible for the activation or repression of associated genes (reviewed in Maniatis et al., 1987). These cis-acting elements appear to mediate high-affinity, sequence specific interactions with trans-acting proteins that determine the transcriptional activity of the gene. The mechanisms by which these proteins influence transcription is still poorly understood (reviewed in Ptashne, 1988), but a preliminary understanding of regulated gene expression and the γ to β switch can be reached by a systematic characterization of cis-acting elements and the factors that bind to them.

Several lines of evidence indicate that the human γ globin promoter is critical to the high level, tissue specific, and developmentally specific expression of the γ globin gene. Nuclease hypersensitive sites appear in

chromatin in the region of both γ globin promoters in
tissues and cell lines that express γ globin (Groudine et
al., 1983; Tuan et al., 1985). These hypersensitive sites
are thought to correspond to nucleosome free regions that
are accessible to transcription factors. DNA mediated gene
transfer studies in transgenic mice and tissue culture cell
lines demonstrate that the γ globin promoter is necessary to
achieve tissue specific and developmentally specific
expression of a linked gene (Trudel et al., 1987; Bodine et
al., submitted; Rutherford and Nienhuis, 1987; Lin et al.,
1987; Donovan-Peluso et al., 1987). Finally, naturally
occurring point mutations within the γ globin promoter have
been shown to alter the developmental specificity of γ
globin gene expression and lead to abnormal persistence of
fetal globin expression into adulthood (reviewed in
Stamatoyannopoulos and Nienhuis, 1987). It has been
suggested that these non-deletion HPFH mutations influence
the binding of transcriptional factors to the promoter
region and mark cis-acting sequences critical to the
developmental specificity of γ globin expression (Collins et
al., 1985; Gelinas et al., 1985).

Prior experiments in our lab suggest sequences within a
fragment of the γ globin promoter from position -259 to -137
may have a role in developmental specificity of γ globin
expression (Lin et al., 1987). This region of the promoter
is interesting in several respects. First, it overlaps with
a nuclease hypersensitive region found in the chromatin of
fetal liver and the K562 human erythroleukemia cell line,
both of which express γ globin and not β globin (Groudine et
al., 1983; Tuan et al., 1985; Gimble et al., 1988). Second,
all but one of the known non-deletion HPFH mutations are
clustered within this fragment (Stamatoyannopoulos and
Nienhuis, 1987). Third, several DNA sequence motifs that
have been demonstrated to bind transcription factors in
other systems are located in this fragment. These include:
a) a CACCC-box motif that has been found to be critical to
the function of the β globin promoter (Orkin et al., 1982;
Charnay et al., 1985; Myers et al., 1986) and that is also
found in the promoters of the human porphobilinogen
deaminase gene (Chretien et al., 1988) and the rat
tryptophan oxidase gene (Schule et al., 1988); b) a
recognition sequence for an erythroid specific factor (Eryfl
or NF-E1) that binds to multiple sites in the chicken and
human β globin 3' enhancers, in the mouse β globin IVS 2,
and in various globin promoters (Evans et al., 1988; Wall et

al., 1988; Galson and Housman, 1988; Mantovani et al.,
1988b; Superti-Furga et al., 1988; Gumucio et al., 1988);
c) an octamer motif that is involved in transcriptional
regulation of the immunoglobulin heavy and light chain genes
(Lenardo et al., 1987; Scheidereit et al., 1987), the
histone H2B gene (Fletcher et al., 1987), the herpes virus
TK gene, snRNA genes, and SV40 early regions (Bohmann et
al., 1987), as well as the replication of adenovirus DNA
(O'Neill et al., 1988). The T to C substitution that occurs
with the -175 HPFH mutation changes the last base in the
octamer sequence.

We have attempted to dissect the upstream region of the
γ globin promoter into its minimal regulatory elements using
a combination of functional and DNA-protein binding assays.
Transient assays have been performed in K562 cells with
truncated and mutated γ globin promoters and the data
correlated with sequences mediating binding of putative
transcriptional factors from erythroid and non-erythroid
cells (McDonagh et al., manuscript in preparation; Lin et
al., manuscript in preparation). The upstream region
appears to be a composite of positive and negative
transcriptional elements and binds several proteins from
both erythroid and non-erythroid cells.

In addition to studies on the promoter, we have
systematically searched the human β globin gene cluster for
enhancer elements that activate the γ globin promoter.
Enhancers located 3' of the chicken and human β globin genes
and an intragenic enhancer in the human β globin gene have
been shown to augment expression of globin promoters in a
variety of experimental models and to correlate with
developmentally specific nuclease hypersensitive sites found
in the chromatin of erythroid cells (Hesse et al., 1986;
Choi and Engel, 1986; Behringer et al., 1987; Antoniou et
al., 1988). Restriction fragments of the β globin locus
were subcloned into a plasmid containing a γ globin promoter
linked to the CAT reporter gene and transient assays were
performed in erythroid and non-erythroid cells. A 751 bp
fragment located 3' to the $^A\gamma$ gene is able to function as an
enhancer of the γ promoter in both erythroid and
non-erythroid cells (Bodine and Ley, 1987).

RESULTS

Gamma Globin Promoter: Correlation between functional
assays, binding assays and HPFH mutations

A series of 5' truncations were made in the upstream γ
promoter from position -374 to -130. These truncation
mutations were linked to a reporter gene and introduced into
K562 cells in a transient assay. No significant change in
activity occured until the promoter was truncated to the Apa
I site at -202, where a 150 to 200 percent increase in
activity was noted compared to the -300 promoter.
Truncation to -163, removing the octamer motif and NF-El
recognition sequence, resulted in a small decrease in
activity. Further truncation to -130, removing the
CACCC-box, results in a fall to 20 percent the activity of
the -300 promoter, marking this region as a site of
significant positive transcriptional activity.

In a complementary set of experiments, the -259 to -137
fragment of the γ promoter was subjected to truncation from
both the 5' and 3' ends. These truncations were linked to a
minimal β globin promoter (-127) and a reporter gene in a
plasmid containing hypersensitive site II from the locus
activating region (Tuan et al., 1985), and transient assays
were performed in K562 cells. The minimal β globin promoter
is transcriptionally silent in this assay. As shown
schematically in Figure 1, these experiments suggest that
the -259 to -137 region is a composite of positive and
negative cis-acting elements. The area around -200, where
the majority of the non-deletion HPFH mutations occur,
appears to be a negative element.

DNAse I footprinting experiments using nuclear extracts
prepared from K562 cells and HeLa cells were performed on a
fragment of the γ promoter from -410 to -53. At least four
regions of protection occured with both K562 and HeLa
extract. These data are summarized schematically in Figure
1. The first protected region is centered at -270 and is
present in both extracts. The second region, also present
in both extracts, is centered at -220 and spans a newly
identified mutation in the $^A\gamma$ promoter associated with
decreased expression of the $^A\gamma$ gene (Gilman et al., 1988).
The third region is centered at -180 and spans both the
octamer motif and the NF-El recognition sequence. The
footprint of this region derived with K562 extract is more

extensive than the footprint with HeLa extract, indicating a
degree of erythroid specificity. The fourth and last
protected region, detected with both extracts, is centered
at -145 and covers the CACCC-box motif. Interestingly, no
protection is observed over the -200 region, where many of
the HPFH point mutations are clustered.

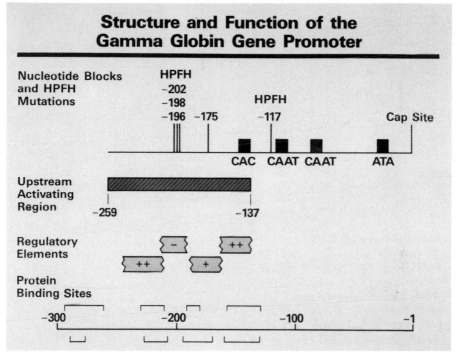

Figure 1: Displayed above is a map of the gamma globin
promoter from -259 to the cap site with the location of
conserved sequence motifs and HPFH point mutations marked.
The location of regulatory elements identified in functional
assays and protein binding sites identified in binding
assays are indicated below.

Gel mobility shift assays were used to further examine
proteins binding to the -180 region and the influence of the
-175 HPFH mutation. Cross competition experiments with an
oligonucleotide containing the wild type octamer region
sequence and an oligonucleotide with the -175 T to C HPFH
mutation reveal that two factors bind to the wild type
sequence while the -175 mutation destabilizes the binding of
one factor. The other complex is unaffected by the -175

mutation. This complex is found in K562 cells and not HeLa cells, suggesting it is erythroid specific and responsible for the broader footprint found with K562 extract. Thus, the -175 HPFH mutation alters binding of a non-erythroid specific protein in a region where an erythroid specific factor also binds.

Mutants were made of the several regions of the γ promoter that had been shown to mediate binding of putative transcriptional factors. These mutated promoters were linked to a reporter gene and tested in transient assays in K562 cells. Mutating the region around -270 has no effect. A mutation which alters the octamer/NF-E1 region results in a slight decrease in activity while a mutation which changes the CACCC-box region reduces activity to 10 percent the activity of the -300 promoter. These findings are consistent with the truncation data and suggest that the CACCC-box confers the majority of the positive transcriptional activity found in the upstream fragment of the promoter in transient assays.

To test if the binding sites themselves contained transcriptional activity independent of other upstream elements, oligonucleotides containing the sequences protected in DNAse I footprinting experiments were cloned individually in front of a γ globin promoter truncated to -130. Both the CACCC-box and octamer regions augment activity in transient assays, consistent with a role as positive regulatory elements. If the -175 HPFH mutation is incorporated into the construct, however, a dramatic 20 fold increase in activity is observed, indicating that this mutation is associated with creation of a powerful positive regulatory element able to activate a minimal γ promoter (McDonagh, manuscript in preparation).

An enhancer is located 3' to the $^A\gamma$ gene

The transient assay system has also been used to identify and characterize a cis-acting element found 3' to the $^A\gamma$ gene that is able to function as an enhancer of the γ globin promoter in erythroid and non-erythroid cells. Restriction fragments of the β globin locus were subcloned into a plasmid containing a γ globin promoter and the CAT reporter gene. As shown in Figure 2, a 751 bp EcoR 1 - Hind III fragment 400 bp downstream of the $^A\gamma$ globin gene polyadenylation site was able to augment CAT activity in transient assays in the erythroleukemia cell lines K562 and HEL as well in HeLa cells (Bodine and Ley, 1987). The enhancer effect is independent of location and orientation. The location of the enhancer corresponds with a developmentally specific nuclease hypersensitive site present in fetal liver and K562 cells. A region of the enhancer 480 bp 3' to the $^A\gamma$ polyadenylation site shows a 17/21 bp match to a region of the chicken β enhancer 410 bp 3' to the β^A polyadenylation site. Further inspection of the enhancer sequence reveals several potential concensus recognition sequences for the erythroid specific transcription factor, NF-E1, and at least one of these

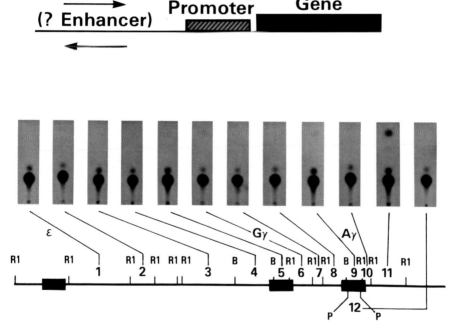

Figure 2: Fragments of the β-globin cluster were cloned in either orientation in front of a γ globin promoter and CAT reporter gene and tested in transient assays in K562 cells. Fragment 11, downstream of the $^A\gamma$ gene, displayed enhancer activity.

sequences appears to bind NF-El in DNAse I footprint experiments and gel mobility shift assays (Purucker et al., unpublished observations). Truncations of the enhancer region are in progress to attempt to subdivide it into minimal functional units.

DISCUSSION

We have characterized two cis-acting elements in the human β globin cluster that appear to be involved in the tissue specific and developmentally specific regulation of human γ globin gene expression. A fragment of the γ globin promoter from -259 to -137 contains regulatory elements able to direct transcription in K562 cells. This fragment binds putative transcriptional factors present in nuclear extracts from erythroid and non-erythroid cells. An enhancer is present 3' to the $^A\gamma$ gene that is able to augment γ promoter activity in both erythroid and non-erythroid cells.

Transient assays in K562 cells demonstrate that the γ promoter upstream of -130 is actually a composite of positive and negative regulatory elements. Such overlapping elements have been observed in the promoters of other genes such as those for interferon (Goodburn and Maniatis, 1988) and human retinol binding protein (Colantuoni et al., 1987). It is not surprising to find these positive and negative elements located in regulatory regions of a gene which undergoes derepression and repression during ontogeny.

The regions around the octamer/NF-El recognition sequences and the CACCC-box appear to contain the majority of the positive transcriptional activity detected in these assays. The wild type octamer region appears to be only weakly positive in spite of the fact that it binds an erythroid specific protein, NF-El, that appears to be an important positive transcriptional factor in other regulatory elements (Evans et al., 1988; Reitman and Felsenfeld, 1988; Wall et al., 1988). The -175 HPFH mutation destabilizes the binding of a non-erythroid specific factor, believed to be OTF-1, and creates a powerful positive regulatory element able to activate a minimal γ promoter independent of other upstream promoter elements (McDonagh et al, manuscript in preparation). Two models have been proposed to explain this effect. The first model suggests that OTF-1 functions as a negative trans-factor that interferes with the positive activity of another factor, presumably NF-El (Mantovani et al., 1988b; Gumucio et al., 1988). The -175 mutation relieves this repression by decreasing the affinity of OTF-1 for this site. OTF-1 has been found to function as a positive or a negative transcriptional factor in different contexts in other systems (Parslow et al., 1987). Interestingly, it has been proposed that NF-El may function as a repressor in the γ promoter when binding to the distal CCAAT box region (Superti-Furga et al., 1988; Mantovani et al., 1988b). The second model proposes that the -175 mutation increases the binding affinity of a positive trans-factor to this area, independent of the effect on OTF-1 (Martin et al., 1988). While the mutations that we have made in this region do not functionally distinguish between these models, it is interesting to observe that several other point mutations within the octamer core recognition sequence have the potential to relieve the OTF-1 mediated repression suggested in the first model, but have not been identified in patients with non-deletion HPFH.

Clinical and experimental observations have indicated that the CACCC-box is critical to the transcriptional activity of the β globin promoter (Orkin et al., 1982; Charnay et al., 1985; Myers et al., 1986). We have observed that the CACCC-box motif is also vital to the full activity of the γ promoter in fetal stage K562 cells. The proximal CACCC-box of the human β globin gene has been shown to bind factors present in both erythroid and non-erythroid cells as well as a factor found only hemin-induced K562 cells and fetal and adult erythroblasts (Mantovani et al., 1988a). The γ globin CACCC-box motif is not able to compete for this erythroid specific factor. We have observed that a minimal β globin promoter, that contains duplicated CACCC-box motifs, is transcriptionally silent in K562 cells. The minimal β promoter can be activated by a small fragment of the γ promoter containing the γ CACCC-box motif (Lin et al., manuscript in preparation). Together these data suggest that the factors binding to the γ and β CACCC-boxes may in fact be different and that the CACCC-box might play a pivotal role in stage specific activation of the globin genes. A more systematic mutational analysis of this region and a careful study of the stage specific factors that bind are warranted to test this hypothesis.

The role of factors binding upstream of -200 is still unclear. The footprint observed at -220 overlaps a recently identified mutation in the $^A\gamma$ gene associated with decreased expression of the $^A\gamma$ gene (Gilman et al., 1988), suggesting that a positive factor may bind in this region. No functional activity can be ascribed to the factor binding at -270 based on our assays. More significantly, no proteins have been observed to bind in the -200 region, in spite of the cluster of HPFH point mutations located in this small area. This may reflect shortcomings in the binding assays that we utilize, or it may suggest that these mutations act indirectly through changes in secondary structure (Tate et al., 1986).

We have also shown that an enhancer is present downstream of the $^A\gamma$ polyadenylation site. This enhancer correlates with a stage specific, erythroid specific nuclease hypersensitive site in chromatin. The enhancer functions in both erythroid and non-erythroid cells, however, when tested in a transient assay system (Bodine and Ley, 1987). There is precedent for a tissue specific enhancer functioning in other tissues, notably the immunoglobulin heavy chain enhancer (Lenardo et al., 1987). The enhancer contains at least one site able to bind NF-E1 in vitro, though it is still unclear if this site is responsible for the enhancing effect seen in erythroid cells (Purucker et al., unpublished observations). NF-E1 has been shown to bind to other globin enhancer elements (Galson and Housman, 1988; Evans et al., 1988; Wall et al., 1988) and to be critical to their activity (Reitman and Felsenfeld, 1988). This factor cannot explain the enhancer activity observed in non-erythroid cells and it is likely that other non-tissue specific trans-factors are involved in the function of this regulatory element.

In correlating the findings of functional experiments and binding assays, a preliminary understanding of the molecular mechanisms regulating γ globin gene expression can be achieved. These models are obviously simplistic in their ability to fully describe the regulated expression this gene. They do not account for the influence of methylation, distant regulatory elements (Grosveld et al., 1987), nuclear matrix (Jarman and Higgs, 1988), and DNA replication (Enver et al., 1988; Epner et al., 1988) on transcription. Nonetheless, they do provide us with testable hypotheses about the mechanisms of hemoglobin switching.

REFERENCES

Antoniou M, deBoer E, Habets G, et al. (1988). The human
β-globin gene contains multiple regulatory regions:
identification of one promoter and two downstream
enhancers. EMBO 7:377-384.
Behringer RR, Hammer RE, Brinster RL, et al. (1987). Two
3' sequences direct adult erythroid-specific expression of
human β-globin genes in transgenic mice. Proc Natl Acad
Sci USA 84:7056-7060.
Bodine DM, Ley TJ (1987). An enhancer element lies 3' to the
human A_γ globin gene. EMBO 6:2997-3004.
Bodine DM, Hoppe PC, Anagnou NP, et al. (1989). Different
patterns of human A_γ globin genes linked to human β globin
genes in MEL cells and transgenic mice. Submitted.
Bohmann D, Keller W, Dale T, et al. (1987). A transcription
factor which binds to the enhancers of SV40,
immunoglobulin heavy chain and U2 snRNA genes. Nature
325:268-272.
Charnay P, Mellon P, Maniatis T (1985). Linker scanning
mutagenesis of the 5'-flanking region of the mouse
β-major-globin gene: sequence requirements for
transcription in erythroid and non-erythroid cells. Mol
Cell Biol 5:1498-1511.
Choi O-R, Engel JD (1986). A 3' enhancer is required for
temporal and tissue-specific transcriptional activation of
the chicken adult β-globin gene. Nature 323:731-734.
Chretien S, Dubart A, Beaupain D, et al. (1988). Alternative
transcription and splicing of the human porphobilinogen
deaminase gene result either in tissue-specific or in
housekeeping expression. Proc Natl Acad Sci USA 85:6-10.
Colantuoni V, Pirozzi A, Blance C, et al. (1987). Negative
control of liver-specific gene expression: cloned human
retinol-binding protein gene is repressed in HeLa cells.
EMBO 6:631-636.
Collins FS, Metherall JE, Yamakawa M, et al. (1985). A point
mutation in the A_γ globin gene promoter in Greek
hereditary persistence of fetal hemoglobin. Nature
313:325-326.
Donovan-Peluso M, Acuto S, Swanson M, et al. (1987).
Expression of human γ-globin genes in human
erythroleukemia (K562) cells. J Biol Chem 262:17051-17057.
Enver T, Brewer AC, Patient RK (1988). Role for DNA
replication in β-globin gene activation. Mol Cell Biol
8:1301-1308.
Epner E, Forrester WC, Groudine M (1988). Asynchronous DNA

replication within the human β-globin gene locus. Proc Natl Acad Sci USA 85:8081-8085.

Evans T, Reitman M, Felsenfeld G (1988). An erythrocyte-specific DNA-binding factor recognizes a regulatory sequence common to all chicken globin genes. Proc Natl Acad Sci USA 85:5976-5980.

Fletcher C, Heintz N, Roeder RG (1987). Purification and characterization of OTF-1, a transcription factor regulating cell cycle expression of a human histone H2b gene. Cell 51:773-781.

Galson DL, Housman DE (1988). Detection of two tissue-specific DNA-binding proteins with affinity for sites in the mouse β-globin intervening sequence 2. Mol Cell Biol 8:381-392.

Gelinas R, Endlich B, Pfeiffer C, et al. (1985). G to A substitution in the distal CCAAT box of the $^{A}\gamma$ globin gene in Greek hereditary persistence of fetal hemoglobin. Nature 313:323-325.

Gilman JG, Johnson ME, Mishima N (1988). Four base-pair DNA deletion in human $^{A}\gamma$ globin-gene promoter associated with low $^{A}\gamma$ expression in adults. Brit J Haem 68:455-458.

Gimble JM, Max EE, Ley TJ (1988). High-resolution analysis of the human γ-globin gene promoter in K562 erythroleukemia cell chromatin. Blood 72:606-612.

Goodbourn S and Maniatis T (1988). Overlapping positive and negative regulatory domains of the human β-interferon gene. Proc Natl Acad Sci USA 85:1447-1451.

Groudine M, Kohwi-Shigematsu T, Gelinas R, et al. (1983). Human fetal to adult hemoglobin switching: changes in chromatin structure of the β-globin gene locus. Proc Natl Acad Sci USA 80:7551-7555.

Grosveld F, van Assendelft GB, Greaves DR, et al. (1987). Position-independent, high-level expression of the human β-globin gene in transgenic mice. Cell 51:975-985.

Gumucio DL, Rood KL, Gray TA, et al. (1988). Nuclear proteins that bind the human γ-globin gene promoter: alterations in binding produced by point mutations associated with hereditary persistence of fetal hemoglobin. Mol Cell Biol 8:5310-5322.

Hesse JE, Nickol JM, Lieber MR, et al. (1986). Regulated gene expression in transfected primary chicken erythrocytes. Proc Natl Acad Sci USA 83:4312-4316.

Jarman AP, Higgs DR (1988). Nuclear scaffold attachment sites in the human globin gene complexes. EMBO 7:3337-3344.

Lenardo M, Pierce JW, Baltimore D (1987). Protein-binding

sites in Ig gene enhancers determine transcriptional activity and inducibility. Science 236:1573-1577.

Lin HJ, Anagnou NP, Rutherford TR, et al. (1987). Activation of the human β-globin promoter in K562 cells by DNA sequences 5' to the fetal γ- or embryonic ζ-globin genes. J Clin Invest 80:374-380.

Maniatis T, Goodbourn S, Fischer JA (1987). Regulation of inducible and tissue-specific gene expression. Science 236:1237-1245.

Mantovani R, Malgaretti N, Nicolis S, et al. (1988). An erythroid specific nuclear factor binding to the proximal CACCC box of the β-globin gene promoter. Nuc Acids Res 16:4299-4313.

Mantovani R, Malgaretti N, Nicolis S, et al. (1988). The effects of HPFH mutations in the human γ-globin promoter on binding of ubiquitous and erythroid specific nuclear factors. Nuc Acids Res 16:7783-7797.

Martin DIK, Tsai S, Orkin SH (1988). An erythroid specific factor binds to the γ globin promoter and mediates expression of an HPFH phenotype. Blood 72:67a.

Myers RM, Tilly K, Maniatis T (1986). Fine structure genetic analysis of a β globin promoter. Science 232:613-618.

O'Neill EA, Fletcher C, Burrow CR, et al. (1988). Transcription factor OTF-1 is functionally identical to the DNA replication factor NF-III. Science 241:1210-1213.

Orkin SH, Kazazian HH, Antonarkis SE, et al. (1982). Linkage of β thalassemia mutations and β globin gene polymorphisms with DNA polymorphisms in human β globin gene cluster. Nature 296: 627-631.

Parslow TG, Jones SD, Bond B, et al. (1987). The immunoglobulin octanucleotide: independent activity and selective interaction with enhancers. Science 235:1498-1501.

Ptashne M (1988). How eukaryotic transcriptional activators work. Nature 335:683-689.

Reitman M, Felsenfeld G (1988). Mutational analysis of the chicken β-globin enhancer reveals two positive-acting domains. Proc Natl Acad Sci USA 85:6267-6271.

Rutherford T and Nienhuis AW (1987). Human globin gene promoter sequences are sufficient for specific expression of a hybrid gene transfected into tissue culture cells. Mol Cell Biol 7:398-402.

Scheidereit C, Heguy A, Roeder RG (1987). Identification and purification of a human lymphoid-specific octamer-binding (OTF-2) that activates transcription of an immunoglobulin promoter in vitro. Cell 51:783-793.

Schule R, Muller M, Otsuka-Murakami H, et al. (1988).
  Cooperativity of the glucocorticoid receptor and the
  CACCC-box binding factor. Nature 332:87-90.
Stamatoyannopoulos G, Nienhuis AW (1987). Hemoglobin
  switching. In Stamatoyannopoulos G, Nienhuis AW, Leder P,
  Majerus P (eds):"The Molecular Basis of Blood Diseases,"
  Philadelphia: WB Saunders, Inc. pp. 66-105.
Superti-Furga G, Barberis A, Schaffner G, et al. (1988).
  The - 117 mutation in Greek HPFH affects the binding of
  three nuclear factors to the CCAAT region of the γ-globin
  gene. EMBO 7:3099-3107.
Tate VE, Wood WG, Weatherall DJ (1986). The British form of
  hereditary persistence of fetal hemoglobin results from a
  single base mutation adjacent to an S1 hypersensitive site
  5'to the $^{A}\gamma$ globin gene. Blood 68:1389-1393.
Trudel M, Magram J, Bruckner L, et al. (1987). Upstream Gγ-
  globin and downstream β-globin sequences required for
  stage-specific expression in transgenic mice.
  Mol Cell Biol 7:4024-4029.
Tuan D, Solomon W, Li Q, et al. (1985). The "β-like-globin"
  gene domain in human erythroid cells. Proc Natl Acad Sci
  USA 82:6384-6388.
Wall L, deBoer E, Grosveld F (1988). The human β-globin gene
  3' enhancer contains multiple binding sites for an
  erythroid-specific protein. Genes and Develop.
  2:1089-1100.

Hemoglobin Switching, Part A: Transcriptional Regulation, pages 163–177
© 1989 Alan R. Liss, Inc.

CIS AND TRANS-ACTING FACTORS REGULATING GAMMA AND BETA GLOBIN GENE EXPRESSION IN K562 CELLS.

Maryann Donovan-Peluso, Santina Acuto, David O'Neill, James Kaysen, Anna Hom, and Arthur Bank

Departments of Genetics and Development, and Medicine, Columbia University, New York, New York, 10032

INTRODUCTION

The K562 cell line is a human erythroleukemia cell line that was derived from a patient with chronic myelogenous leukemia (Lozzio and Lozzio, 1975). These cells express the endogenous epsilon and gamma genes and do not express the adult beta globin genes (Rutherford T, et. al., 1981). The lack of beta globin gene expression occurs despite the fact that these cells contain an intact beta gene that is capable of being expressed in transient expression assays (Donovan-Peluso et. al. 1984; Fordis et. al., 1984). In addition transfected epsilon genes are expressed and transfected beta genes are not expressed (Young et. al., 1984; Young et. al., 1985). When the cells are grown in hemin, induction, or an increase in the transcription and the accumulation of globin gene transcripts occurs (Charnay P., Maniatis T, 1983; Miller CW, et. al. 1984). Thus, K562 cells represent the embryonic-fetal stage of erythroid maturation and provide a model system for studying the regulation of gamma globin gene expression.

RESULTS

We have used several approaches to characterize the cis-acting sequences and trans-acting factors involved in the regulation of gamma globin gene expression. Initial studies have examined the expression of fusion genes comprised of gamma and beta gene elements to define the important

sequences required for expression and induction
(Donovan-Peluso M, 1987). These genes are
schematized in Figure 1.

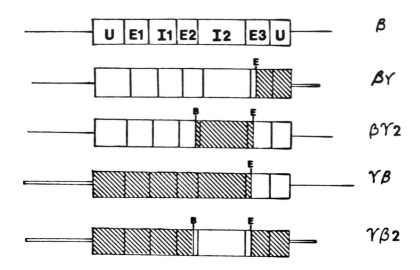

Figure 1.   Fusion genes analyzed in K562 cells.
The top line shows the beta globin gene with the
5' and 3' untranslated regions (U), three exons
(E1,E2,E3) and two introns (I1,I2). Beta elements
are indicated by single lines and open boxes and
gamma elements by double lines and hatched boxes.

        The beta gene is contained on a Pst I frag-
ment of 4.4 Kb. The beta-gamma gene contains beta
5' and structural gene sequences to the Eco RI
site in exon 3 and gamma sequences 3'. Beta-gamma
2 is a beta gene that contains the substitution of
a Bam HI to Eco RI fragment containing gamma IVS 2
for the identical beta fragment containing beta
IVS 2. Gamma-beta contains gamma 5' and structural gene sequences to the Eco RI site in exon 3
and beta sequences 3' including the beta 3' enhancer element. Gamma beta 2 contains the gamma gene
as an Eco RI 3.2 Kb fragment into which the Bam
HI-Eco RI fragment containing beta IVS 2 has been
substituted for the identical fragment containing
gamma IVS 2.

Table 1 summarizes the effect of promoter, IVS2, and beta 3' enhancer on fusion globin gene expression and induction. The beta gene containing the beta promoter, beta IVS2, and no 3' enhancer element is neither expressed nor induced as indicated by the negative signs in the two columns at the right. The substitution of the gamma 3' region for the beta 3' region in the beta-gamma fusion gene does not allow this gene to be expressed suggesting that no critical element for gamma expression resides in this region. The substitution of gamma IVS2 for beta IVS2 in beta-gamma2 generates a fusion gene that is capable of being expressed and induced when transcripts are analyzed with a 3' probe, however, these transcripts do not originate at the canonical beta globin cap site and in fact, even by primer extension no discrete 5' ends are detectable. The expression of this gene suggests that gamma IVS2 may play a role in the expression and induction of gamma genes in K562 cells but this element is not sufficient to promote proper initiation of transcription of a gene containing a beta promoter in K562 cells. With the gamma-beta fusion gene containing a gamma promoter, gamma IVS2, and beta 3' enhancer significant expression and induction is observed in 50% of the lines. These transcripts are appropriately initiated as shown by primer extension analysis. The gamma-beta2 fusion gene containing the gamma promoter, beta IVS2, and gamma 3' region is expressed but induction is observed in only 1 of 6 clones analyzed. The lack of inducibility of this gene may be an effect of either the substitution of beta IVS2 for gamma IVS2 or the lack of the beta 3' enhancer element on this construct.

		PROMOTER	IVS2	3' ENHANCER	EXP	IND
1.	β	β	β	NO	−	−
2.	βγ	β	β	NO	−	−
3.	βγ2	β	γ	NO	+	+
4.	γβ	γ	γ	YES	+	+
5.	γβ2	γ	β	NO	+	−

Table 1. Summary of the effect of promoter, IVS 2, and beta 3' enhancer on fusion globin gene expression and induction.

To further define the role of the gamma promoter, IVS2 and the beta 3' enhancer on expression
and induction three additional genes were constructed. Figure 2 contains a schematic representation of these genes. Each of these genes contains the gamma promoter fused to the beta structural gene sequences at an NcoI site at the translational AUG. In addition, the first two fusion
genes (gamma promoter beta and gamma promoter beta
delta 3') contain beta IVS2. As indicated by the
longer single line at the 3' end of gamma
promoter-beta the beta 3' enhancer is included on
this construct. The next gene, gamma promoter-
beta delta 3', does not contain the beta 3' enhancer. The final fusion gene gamma promoter-beta
gamma 2 delta 3' contains the substitution of
gamma IVS2 for beta IVS2 and does not contain the
beta 3' enhancer.

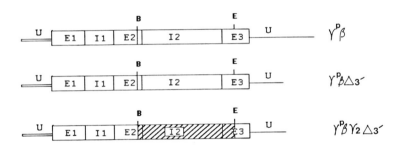

Figure 2. Schematic representation of additional
fusion genes transfected into K562 cells. Gamma
elements are indicated by double lines and hatched
boxes and the beta elements by single lines and
open boxes.

Figures 3-5 contain the results of Rnase protection analysis of transcripts produced by the
fusion genes diagrammed in Figure 2. Figure 3
shows an analysis of pools of clones transfected
with gamma promoter-beta. The fusion probe complementary to the 5' end of this RNA is depicted
below and the protected fragment of 146
nucleotides is indicated. Lane M contains the
marker, lane P contains the probe. Lanes P1 and
P2 contain RNA from pools of clones transfected
with gamma promoter beta and either uninduced (-)

or induced (+) with hemin. As can be seen this gene containing the gamma promoter, beta IVS 2 and the beta 3' enhancer is both expressed and induced. The gamma promoter is sufficient to allow expression of a gene in K562 cells.

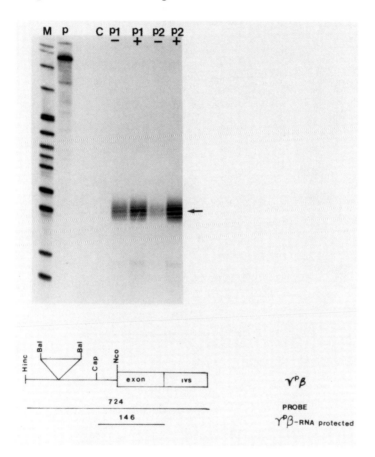

Figure 3. Rnase protection analysis of pools of clones transfected with gamma promoter beta.

Figure 4 shows the analysis of pools of cells transfected with gamma promoter beta without the 3'beta enhancer. The 5' fusion probe that was used is schematized at the bottom and the pro- tected fragment of 146 nucleotides is indicated. Lane M contains the marker, c contains non- transfected K562 cell RNA and p1, p2 and p3 con-

tain RNA from pools of transfected cells. Again
(-) indicates uninduced cells and (+) indicates
induced cells. The arrow at the right indicates
the protected fragment of 146 nucleotides. The
expression of this gene lacking the beta 3' enhancer and containing beta IVS2 decreases with induction consistent with a role for the beta 3' enhancer on induction.

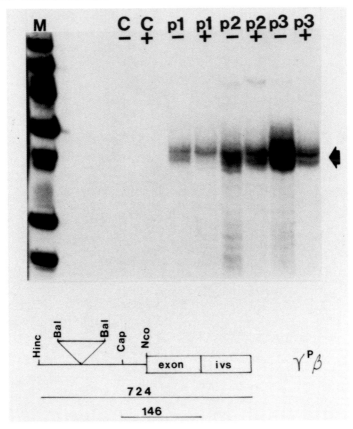

Figure 4. Rnase protection analysis of pools of
clones transfected with gamma promoter beta delta
3'.

Figure 5 shows an analysis of 10 individual
clones transfected with gamma promoter beta gamma
2 delta 3' a gene that contains gamma IVS 2 but
does not contain the beta 3' enhancer. The numbers 1-10 indicate the ten individual clones, (-)

indicates uninduced cells and (+) indicates
induced cells. The arrow at the right indicates
the protected fragment of 146 nucleotides. As
seen in lanes 2,3,4,5, and 9 this gene is both
expressed and induced. In total, approximately
50% of the clones are able to increase the
expression of this gene in response to hemin
induction. These data are consistent with a role
for gamma IVS2 in the induction of globin genes in
K562 cells even in the absence of the 3' beta
enhancer. The relative roles of the 3' beta
enhancer and gamma IVS 2 on the <u>in vivo</u> expression
of the gamma gene remains to be determined.

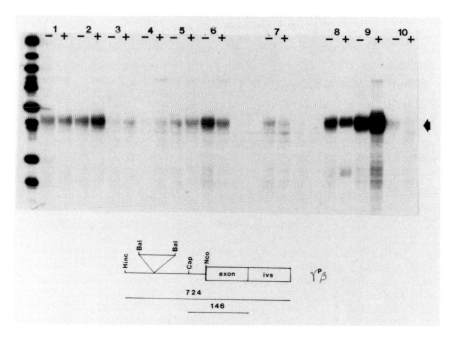

Figure 5. Rnase protection analysis of individual
clones transfected with gamma promoter beta gamma
2 delta 3'.

To further define the role of the beta 3'
enhancer on induction and to define which other
sequences within the structural globin gene are
required to interact with this element transcription and induction of gamma-neo fusion genes shown
in Figure 6 were evaluated. The gamma promoter to

an AcyI site in the 5' untranslated region indicated by the hatched box was fused to the structural neomycin resistance gene indicated by the
double lines. The beta 3' enhancer contained
within an EcoRI-Bgl II fragment was ligated
downstream of the gamma-neo gene in both orientations. As reported previously (Acuto S, et. al.,
1987) the gamma-neo gene alone was expressed but
not induced.

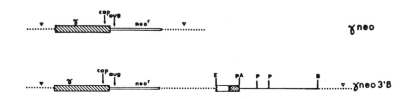

Figure 6. Gamma-neo fusion genes introduced into
K562 cells. Gamma elements are indicated by
hatched boxes, neomycin resistance gene elements
are indicated by open boxes, and beta gene elements are indicated by dark lines and stippled
boxes.

Figure 7 shows the results of an Rnase protection analysis of K562 cells transfected with
gamma-neo plus the 3' beta enhancer. RNA from
individual clones is contained in lanes numbered
1-4 and RNA from one pool of clones (labelled p).
RNA from uninduced (-) and induced (+) cells was
hybridized to the fusion probe shown at the bottom
of the slide. The protected fragment of 265
nucleotides is indicated by the top arrow at the
right. The lower arrow at the right indicates the
actin internal control signal. This signal is
used to control for the amount of mRNA present in
each lane and should be consistent between the
uninduced and induced lanes for each sample.
After adjusting for variations in the concentration of mRNA in the uninduced and induced lanes by
using the actin signal one can see that the gamma-
neo plus the 3' beta enhancer induces in response
to hemin induction. These data provide further
evidence that the beta 3' enhancer is involved in
induction in K562 cells. The beta 3' element does

not appear to be gene specific and does not seem
to require the interaction with sequences within
the structural globin genes for its effect.
Although this element may be cell type specific it
does not appear to be stage specific since it is

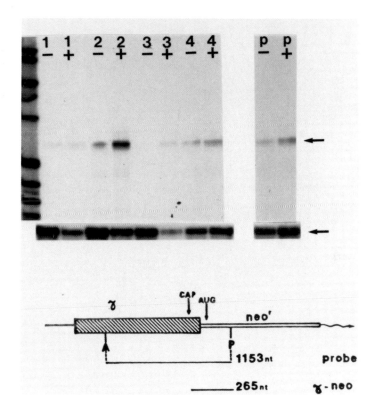

Figure 7. Rnase protection analysis of RNA from
cells transfected with gamma-neo plus the 3' beta
enhancer element.

functional in an embryonic-fetal erythroid
environment. This element is able to act independently of gamma IVS2 a second element that
appears to be involved in induciblilty in K562
cells. Whether or not gamma IVS 2 is developmental stage specific remains to be evaluated.

Figure 8. Schematic representation of the gamma
5' region.

     In addition to the analysis of fusion gene
transcription we have also used the gel mobility
shift assay to identify the presence of factors in
extracts from erythroid and nonerythroid cells
Kaysen J et. al., 1989). Figure 8 shows a
schematic representation of the gamma 5' region.
The CAP, TATA, and CAAT boxes are indicated. The
asterisks indicate the mutations associated with
HPFH. The restriction sites used to generate
probes for binding are shown.

     Figure 9 shows the results of binding to a 42
basepair Hae III-Ava II fragment containing the
gamma globin octamer sequence. Lane p is the
probe alone, lane H contains probe incubated with
extract from Hela cells a nonerythroid control,
lane K- contains probe incubated with uninduced
K562 cell extract and lane K+ contains probe
incubated with induced K562 cell extract. The
arrows at the right indicate the bands with
altered mobility. Band A is present in both
erythroid and nonerythroid cell extracts. Band B
is present in erythroid extracts and the intensity
does not seem to change with induction. Band C is
seen only in erythroid cell extracts and not in
HeLa cells. This band seems to diminish in
intensity with induction suggesting that either
some factor is less abundant or has altered binding affinity to the gamma42 region in induced K562
cells.

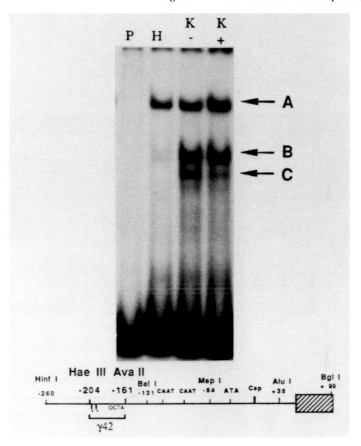

Figure 9. Gel mobility shift assay using extracts from K562 cells to bind with the gamma 5' 42 basepair Hae III-Ava II fragment.

Figure 10 shows the results of a competition analysis to show that the three bands with altered mobility result from specific binding to the gamma42 fragment. The three bands A, B, and C are indicated at the right. Lane 1 contains the probe alone, lane 2 contains probe incubated with K562 extract, lanes 3 and 4 contain probe incubated with K562 extract to which two levels of unlabelled gamma42 fragment were added as specific competitor, lanes 5 and 6 contain probe incubated with K562 extract to which nonspecific competitor was added. Only addition of the unlabelled gamma42 fragment is able to compete for binding of factors showing that the binding is specific.

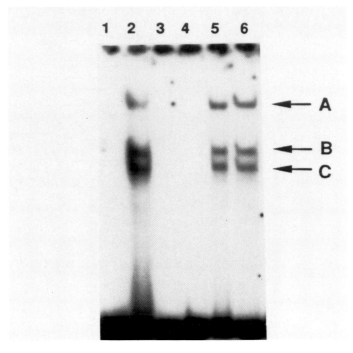

Figure 10. Competition analysis to show that the three bands with altered mobility result from specific binding to the gamma42 fragment.

In order to determine whether or not the binding of factors to the octamer sequence was erythroid cell specific and changed following induction a double stranded oligonucleotide containing the octamer sequence was synthesized and used as competitor. Figure 11 shows the results of this analysis. Lane 1 contains probe alone; lane 2 contains probe incubated with K562 extract; lanes 3-5 contain probe incubated with K562 cell extract to which increasing amounts of double stranded DNA containing the octamer sequence has been added as competitor. The octamer competitor DNA abolishes the binding to the A band. This band is not erythroid specific and does not change with induction. Band C, the one that changes with induction, does not change when the octamer sequence is added as competitor suggesting that

binding of factors to some other sequence on the
gamma42 fragment is involved in the altered binding seen in extracts from induced K562 cells.

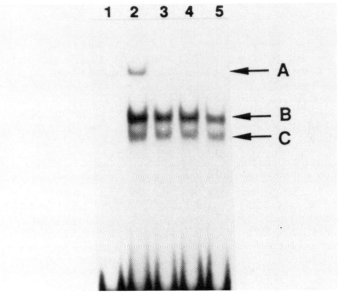

Figure 11. Competition analysis with an octamer
oligonucleotide to show which shifted bands result
from binding of the octomer specific factors to
the gamma 42 fragment.

In summary, these experiments have shown that
the presence of 5' gamma flanking sequences on a
beta gene (gamma promoter beta) is sufficient for
its expression and induction in K562 cells. Gamma
IVS 2 plays a role in gamma globin gene expression
and induction. The 3' beta flanking enhancer
sequences may influence the induced expression of
gamma-beta and gamma-neo fusion genes. The -225
to -183 (gamma 42) gamma 5' flanking region contains sequences that bind both erythroid cell
specific and non-erythroid cell factors. The
binding of a K562 specific factor to gamma42
decreases with erythroid maturation. The binding
of factors to the octamer sequence contained in
the gamma42 fragment is not erythroid specific and
does not change with induction. These results
indicate that multiple genetic elements are
involved in the regulation of gamma globin gene
expression in K562 cells.

REFERENCES

Acuto S, Donovan-Peluso M, Giambona N, Bank A
(1987). The role of human globin gene
promoters in the expression of hybrid genes in
erythroid and non-erythroid cells. Biochem.
Biophys. Res. Commun. 143:1099-1106.

Charnay P, Maniatis T (1983). Transcriptional
regulation of globin gene expression in the
human erythroid cell line K562. Science 220:
1281-1283.

Donovan-Peluso M, Young Katherine, Dobkin C,
Bank A (1984). Erythroleukemia (K562) cells
contain a functional beta globin gene. Mol. and
Cell. Biol. 4(11): 2553-2555.

Donovan-Peluso M, Acuto S, Swanson M, Dobkin C,
Bank A (1987). Expression of human gamma globin
genes in human erythroleukemia (K562) cells.
The J. of Biol. Chem. 262(15): 17051-17057.

Donovan-Peluso M, Acuto S, Bank A (1987). Regulation of fetal globin gene expression in
human erythroleukemia (K562) cells. In
Molecular Biology of Hemopoiesis. M Tav-
assleoli, NG Abraham, JL Ascensao, ED Zanjani,
and A Levine, Eds. Plenum Publishing, New
York, NY.

Fordis CM, Anagnou NP, Dean A, Nienhuis AW,
Schecter AN (1984). A beta globin gene
inactive in the K562 leukemic cell, functions
normally in a heterologous expression system.
Proc. Natl. Acad. Sci. 81:4485-4489.

Kaysen J, Donovan-Peluso M, Acuto S, O'Neill D,
Bank A (1989). Regulation of human fetal
hemoglobin gene expression. Annals of the New
York Acad. Sci. In press.

Lozzio CB, Lozzio BB (1975). Human chronic
myelogenous leukemia cell line with positive
philadelphia chromosome. Blood 45: 321-324.

Miller C, Young K, Dumenil D, Alter BP, Schofield
JM, Bank A (1984). Specific globin mRNAs in
human globin genes in human erythroleukemia
(K562) cells. Proc. Natl. Acad. Sci. 81:5315-
5319.

Rutherford T., Clegg JB, Higgs DR, Jones RW,
Thompson J., Weatherall DJ (1981). Embryonic
erythroid differentiation in the human leukemic

cell line K562. Proc. Natl. Acad. Sci. 78(1):
348-352.
Young K, Donovan-Peluso M, Bloom K, Allan M, Paul
J, Bank A (1984). Stable transfer and
expression of exogenous human globin genes in
human erythroleukemia (K562) cells. Proc. Natl.
Acad. Sci. 81:5315-5319.
Young K, Donovan-Peluso M, Cubbon R, Bank A (1985)
Trans-acting regulation of beta globin gene
expression in erythroleukemia (K562) cells.
Nucl. Acids Res. 13(14):5203-5213.

Hemoglobin Switching, Part A: Transcriptional Regulation, pages 179–191

BINDING OF AN ERYTHROID-SPECIFIC FACTOR TO ENHANCER REGIONS OF HUMAN GLOBIN GENES

Qi-Hui Gong and Ann Dean

Laboratory of Cellular and Developmental Biology, NIDDK, NIH, Bethesda, MD 20892

INTRODUCTION

A fundamental question in developmental biology is how gene expression is controlled in a temporal and tissue-specific manner. The individual genes of the human α- and β-globin clusters are temporally regulated to bring about the sequential production of embryonic, fetal, and adult hemoglobins during ontogeny (Karlsson and Neinhuis, 1985). The mechanisms controlling these developmental switches are complex and not well understood.

We are interested in the changes in chromatin structure that genes undergo when they are activated during development. Structural studies have shown the appearance of DNase I hypersensitive sites in and near the human globin genes when these genes are expressed (Groudine et al., 1983; Tuan et al., 1985; Forrester et al., 1986). In addition, nucleosome positions are altered when the β-major globin gene of mouse is activated (Benezra et al., 1986). Sequence-specific DNA binding proteins are likely to be involved in these structural changes (Emerson et al., 1985). A detailed understanding of how protein-DNA complex formation might affect the structure of chromatin and gene transcription (either positively or negatively) is currently the subject of considerable interest.

Recently, an erythroid-specific protein has been described which binds to two adjacent sites in the chicken β-globin enhancer (Evans et al., 1988). The protein recognizes the hexamer consensus sequence A/TGATAA/G, and has additional binding sites within regulatory regions of all chicken globin genes. The specificity of this factor is similar, if not identical, to factors which have binding sites in the human γ-globin promoter (Mantovani et al., 1987), the mouse β-globin IVS 2 region (Galson and Housman, 1988), and the human β-globin enhancer (Wall et al., 1988).

We now describe an erythroid-specific factor in nuclear extracts of K562 cells which binds to one of two tandem (and inverted) copies of the chicken hexamer consensus sequence which are present in the 3' flanking region of the human ϵ-globin gene. By competition studies we show that the same activity binds to two sites in the $^A\gamma$-globin enhancer, and to at least one site in the β-globin enhancer. The ϵ-globin 3' binding site for this factor is in a region where several tissue and developmental stage-specific DNase I hypersensitive sites have been reported (Zhu et al., 1984) but where regulatory functions are yet to be described.

RESULTS AND DISCUSSION

Gel mobility-shift assays (Fried and Crothers, 1981; Garner and Revzin, 1981) were used to study the binding of human erythroid nuclear factors to chicken and human globin gene sequences. Oligonucleotide probes were incubated with nuclear extracts from induced K562 cells. These cells actively transcribe the globin genes, with the exception of the β-globin gene (Charney and Maniatis, 1983; Dean et al., 1983), and might be expected to contain positive regulatory factors for these genes. Figure 1A shows that nuclear extracts from induced K562 cells contain a binding activity with high affinity for an oligo-

TABLE 1. Sequences of Oligonucleotides Used in Binding Studies

---

Chicken βL	aattcGGTTGC<u>AGATAAA</u>CATTTg*
ϵ11	aattcTTTTCAGCTC<u>TGATAA</u>C<u>TATCA</u>TTCTACTCTCAg
ϵ11 L	aattcTTTTCAGCTC<u>TGATAA</u>g
ϵ11 R	aatt<u>CTATCA</u>TTCTACTCTCAg
ϵ11 ML	aattcTTTTCAGCTCTTCGCA<u>CTATCA</u>TTCTACTCTCAg
ϵ11 MR	aattcTTTTCAGCTC<u>TGATAA</u>CGCGAATTCTACTCTCAg
γ1	aattCTGACC<u>TTATCT</u>GTGGGG
γ2	aattcTTTCC<u>TTATCA</u>GAAGCg
β1	aattcGGAA<u>CTATCA</u>CTCTTg
β3	aattCTCCC<u>TTATCA</u>TGTCCg

---

* Oligonucleotides were synthesized on a Biosearch 8700 Synthesizer and purified on 15% acrylamide/8M urea sequencing gels. Upper and lower strands were annealed to form double stranded molecules; only the sequence of the upper strand is shown. The hexamer consensus sequence is underlined. Capital letters indicate nucleotides present in the genomic DNA sequence.Oligonucleotides probes were labelled on the 5'end using T4 kinase and [^{32}P]-γATP prior to annealing with the unlabelled complementary strand.

nucleotide representing the sequence (see Table 1) of the left side of region IV of the chicken β-globin enhancer (Emerson et al., 1987). The chicken βL probe (Figure 1A, lanes 1-5) contains a single copy of the chicken hexamer consensus sequence which is duplicated at its genomic locus.

Next, an oligonucleotide representing a region (ϵ11) located about 3.5kb downstream of the human ϵ-globin gene cap site (see Table 1 for

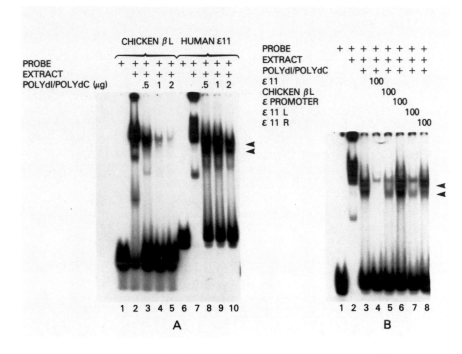

Figure 1.Gel mobility shift assays comparing complex formation between induced K562 cell nuclear
extracts and either the chicken βL probe or the
human ε11 probe (panel A). In panel B various
unlabelled DNAs (see text) are compared for their
ability to disrupt complex formation between induced K562 cell nuclear proteins and probe ε11.
Nuclear extracts were prepared from tissue culture cells as described (Dignam et al., 1983),
and contained between 5 and 15mg of protein per
ml. Assay mixtures (20μl) contained 0.3ng of
labelled oligonucleotide probe, 0.5-2μg poly
dI/poly dC, 100ng of competitor as indicated, and
10-15μg of nuclear protein extract in 10mM Tris
pH 7.5/50mM NaCl/1mM DTT/1mM EDTA/5% glycerol.
Mixtures were incubated for 10 min at 4°C, and
separated on 6% acrylamide gels (29:1 acryla-
mide:N,N'-methylenebisacrylamide) run in TBE at
200 V. Gels were dried before autoradiography.

sequences) was used as a probe. This region contains two copies of the hexamer sequence.
In the presence of excess non-specific competitor DNA (poly dI/poly dC) two major shifted bands were observed (Figure 1A, lanes 8-10), suggesting the presence of one or more binding activities with high affinity for this sequence. In experiments with diminishing amounts of nuclear extract both bands showed a similar decrease in intensity (not shown) consistent with the presence of two different binding activities. Extracts from uninduced cells gave the same results.

To test the specificity of the DNA-protein complex, gel mobility shift assays using probe ε11 were performed in the presence of excess unlabelled oligonucleotide. Addition of a 300-fold molar excess of an oligonucleotide representing a portion of the human ε-globin promoter which contains no sequence similarities with the ε11 region, had no effect on the formation of the complexes (Figure 1B, lane 6). However, the ε11 oligomer was able to compete with itself, and the chicken βL oligomer competed for formation of the major (more slowly migrating) complex (Figure 1B, lanes 4 and 5).

The latter result suggests that the binding activity in K562 nuclear extracts which recognizes a sequence in both the chicken βL oligonucleotide and the ε11 region has only one binding site on the ε11 oligonucleotide even though two hexamer sequences are present. When oligonucleotides (see Table 1) containing exclusively sequences from either the 5' (ε11L), or the 3' (ε11R) portions of ε11 were tested for their ability to compete for factor binding, only ε11L competed efficiently (Figure 1B, lanes 7 and 8) suggesting that the more 5' hexamer sequence in ε11 binds the factor which also recognizes the chicken βL oligomer.

The oligomer containing exclusively the 5' hexamer sequence of ε11 (ε11L) was tested for its ability to form complexes with K562 nuclear

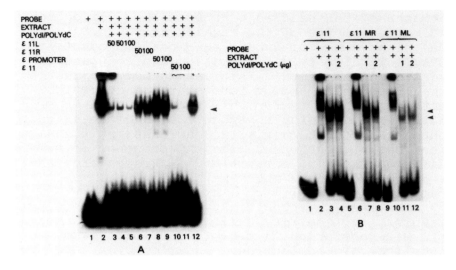

Figure 2. Gel mobility shift was used to study complex formation between the ∈11L probe and nuclear proteins from induced K562 cells(panel A). Amounts of competitor DNAs are in nanograms. In panel B the probe DNA is indicated. Details of the experiment are the same as given in the legend to Figure 1.

extracts. In gel retardation experiments using extracts from uninduced K562 cells a single complex was formed (Figure 2A, lane 12) which could be competed efficiently by itself (Figure 2A, lanes 3-5), and by ∈11 (Figure 2A, lanes 10-11). An oligonucleotide containing the 3' hexamer sequence of ∈11 (∈11R) was a much weaker competitor (Figure 2A, lanes 6 and 7), as was a non-specific oligonucleotide from the ∈-globin promoter region (Figure 2A, lanes 8 and 9).

The potential binding sites in oligonucleotide ∈11 were tested directly by introducing mutations into the sequence. Gel mobility shift experiments were performed using extracts from induced K562 cells and probes that contained 4-bp mutations in the 5' (∈11ML) or the 3' (∈11MR) hexamer sequences present in ε11 (see Table 1 for sequences). When the four central nucleotides of the 5' hexamer sequence were mutated, formation of the major complex was eliminated (Figure 2B, compare lanes 3 and 4 with lanes 11 and 12). Formation of the minor, more rapidly moving complex was still observed. When the mutations were introduced into the center of the 3' hexamer sequence the major complex was still capable of forming, and the minor complex formed only at diminished levels (Figure 2B, lanes 7 and 8).

Based on these data, and the mobility shift data of lanes 7 and 8 of Figure 1B, it is likely that a factor in K562 nuclear extracts recognizes and binds to only one of the hexamer sequences in ε11, the more 5' sequence. A second activity may bind with a different specificity to the sequences in the 3' portion of ε11 since it still has some ability to form a complex when the central four nucleotides of the 3' hexamer are mutated. Alternatively, it is possible that there is only a single hexamer recognizing activity in K562 nuclear extracts which binds less tightly to the 3' hexamer sequence of ε11 because the 3' hexamer lacks other (unidentified) sequence elements which contribute to the strength of interaction. Indeed, within the region encompassing 2 kilobases downstream of the ε-globin gene polyA addition site the hexamer sequence appears 11 times. However, oligonucleotides representing these regions differed greatly in their ability to compete with ε11 (data not shown).

We looked for a binding activity specific for the region contained in oligonucleotide ε11 in nuclear extracts from non-erythroid cells. Gel mobility shift experiments with a HeLa cell extract (Figure 3A, lane 5) and an extract of A431 cells (a human epidermal carcinoma cell line, Figure 3A, lane 7) showed only the minor complex (with A431 cells this is barely detectable). In contrast, extracts from both induced and uninduced K562 cells (Figure 3A, lanes 3 and 4) gave rise to the same major and minor complexes previ-

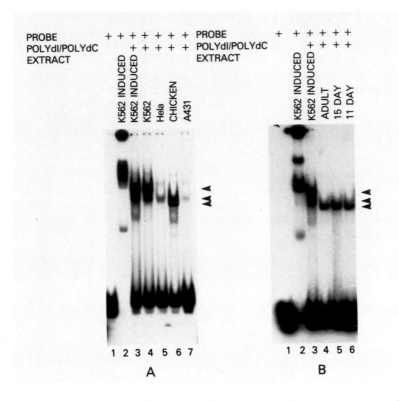

Figure 3. Gel mobility shift experiment comparing the interaction of probe ε11 with extracts of nuclear proteins from various erythroid and non-erythroid cells. Details of the experiment are the same as in the legend to Figure 1. Similar amounts of protein extract were used in each lane. Nuclear extracts of chicken erythrocytes at various developmental stages were a gift of T. Evans and G. Felsenfeld. A431 cell nuclear extracts were a gift of A. Johnson and G. Merlino.

ously seen. The results, taken together with the data presented in Figures 1 and 2 suggest that there are indeed two binding activities for ε11 in K562 nuclear extracts: one is erythroid cell-specific and the other is not.

An extract from nuclei of chicken erythroid cells (from 15 day embryos) formed a single complex with probe ϵ11 which migrated more rapidly than the complex formed with the K562 extract (Figure 3A, lane 6). This complex was formed with extracts from different chicken developmental stages (Figure 3B, lanes 4-6). These data, and the data of Figure 1A support the idea that the human and chicken nuclear binding activities recognize the same sequence element in the DNA of the two organisms.

Two hexamer consensus sequences are present in the minimal $^A\gamma$-globin 3' enhancer (Bodine and Ley, 1987). These appear at positions +2411 (γ1) and +2493 (γ2) with respect to the $^A\gamma$-globin cap site. Interestingly, neither sequence appears in the corresponding region of the $^G\gamma$-globin gene. Each of these sequences was placed at the center of a synthetic oligonucleotide (see Table 1 for sequences) and used as a probe in gel mobility shift experiments with a K562 nuclear extract. Figure 4A shows that each of these probes forms a single high affinity complex with a binding activity in the K562 nuclear extract.

In Figure 4B various oligonucleotides (at about 300-fold molar excess) were tested as competitors of complex formation between the γ1 probe and the K562 binding activity. The probe itself, as well as γ2, ϵ11 and the chicken βL oligonucleotides prevented complex formation (Figure 4B, lanes 2,3,6 and 8 respectively). Specific complex formation could not be competed by an unrelated oligonucleotide representing an ϵ-globin promoter sequence (Figure 4B, lane 7).

Also tested as competitors were oligonucleotides representing sequences 3' of the human β-globin gene. Three hexamer sequences appear in the minimal β-globin 3' enhancer (Wright et al., 1984; Behringer et al., 1987; Kollias et al., 1987; Trudel et al., 1987). These appear at positions +2181 (β1), +2297 (β2), and +2390 (β3). The hexamer sequences of β1 and β3 were synthe-

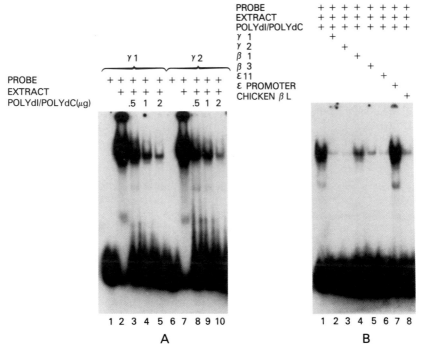

Figure 4. Gel mobility shift experiment showing the interaction between the γ1 and γ2 oligonucleotide probes and a K562 nuclear extract (panel A). In panel B various oligonucleotides containing sequences from regions of the 3' β-, and γ-globin enhancer regions are compared for their ability to prevent complex formation between probe γ1 and K562 nuclear proteins. Details of the experiment are the same as described in the legend to Figure 1.

sized as part of oligonucleotides (see Table 1 for sequences) and used as competitors. β3 was a much stronger competitor of complex formation between the γ1 probe and the K562 binding activity than was β1 (lanes 4 and 5). This experiment suggests that a single K562 nuclear factor, which has at least one strong binding site in the 3' region of the ε-globin gene, also binds strongly to two sites in the $^A\gamma$-globin 3' enhancer, and at least one site in the β-globin 3' enhancer.

CONCLUSIONS

An erythroid-specific factor is present in nuclear extracts of induced and uninduced K562 cells which has strong specific binding sites in the 3' flanking regions of the human ϵ-, $^A\gamma$-, and β-globin genes. The recognition sequence of this binding activity is a hexamer, A/TGATAA/G, which is an element in the chicken β-globin enhancer (and which appears in regulatory regions of all other chicken globin genes) for which a binding activity in chicken erythroid cells has been described (Evans et al., 1988). Since the factor binds (at least in vitro) to genes expressed at different developmental stages, it may be a general factor involved in the regulation of globin gene expression. Alternatively, developmental stage-specific expression could be achieved in concert with other factors or by template availability. The binding sites for the factor in the $^A\gamma$- and β-globin genes are in regions which have enhancer activities. The functional involvement of the ϵ-globin gene binding site in gene regulation remains to be investigated.

REFERENCES

Behringer RR, Hammer RE, Brinster RL, Palmiter RD, Townes TM (1987). Two 3' sequences direct adult erythroid-specific expression of human β-globin genes in transgenic mice. Proc Nat Acad Sci USA 84:7056-7060.

Benezra R, Cantor CR, Axel R (1986). Nucleosomes are phased along the mouse β-major globin gene in erythroid and non-erythroid cells. Cell 44:697-704.

Bodine D, Ley TJ (1987). An enhancer element lies 3' to the human $^A\gamma$ globin gene. EMBO J 6:2997-3004.

Charney P, Maniatis T (1983). Transcriptional regulation of globin gene regulation in the human erythroid cell line K562. Science 220:1281-1283.

Dean A, Ley TJ, Humphries RK, Fordis M, Schechter AN (1983). Inducible transcription of five globin genes in K562 human leukemia cells. Proc Nat Acad Sci USA 80: 5515-5519.

Dignam JD, Lebovitz RM, Roeder RG (1983). Accurate transcription initiation by RNA polymerase II in a soluble extract from isolated mammalian nuclei. Nucleic Acids Res 11:1475-1489.

Emerson B, Lewis CD, Felsenfeld G (1985). Interaction of specific nuclear factors with the nuclease-hypersensitive region of the chicken adult β-globin gene: Nature of the binding domain. Cell 41:21-30.

Emerson B, Nickol J, Jackson PD, Felsenfeld G (1987). Analysis of the tissue-specific enhancer at the 3' end of the chicken adult β-globin gene. Proc Nat Acad Sci USA 84: 4786-4790.

Evans T, Reitman M, Felsenfeld G (1988). An erythrocyte-specific binding factor recognizes a regulatory sequence common to all chicken globin genes. Proc Nat Acad Sci USA 55:5976-5980.

Forrester WC, Thompson C, Elder JT, Groudine M (1986). A developmentally stable chromatin structure in the human β-globin cluster. Proc Nat Acad Sci USA 83:1359-1386.

Fried M, Crothers DM (1981). Equilibria and kinetics of lac repressor-operator interactions by polyacrylamide gel electrophoresis. Nucleic Acids Res 9:6505-6525.

Galson D, Housman D (1988). Detection of two tissue-specific DNA binding proteins with affinity for the mouse β-globin intervening sequence 2. Mol Cell Biol 8: 381-392.

Garner MM, Revzin A (1981). A gel electrophoresis method for quantifying the binding of proteins to specific DNA regions: Application to components of the Escherichia coli lactose operon regulatory system. Nucleic Acids Res 9: 3047-3060.

Groudine M, Kohwi-Shigematsu T, Gelinas R, Stamatoyannopoulos G, Papyannopoulou T (1983). Human fetal to adult globin switching: Changes in chromatin structure of the β-globin locus. Proc Nat Acad Sci USA 80:7551-7555.

Karlsson, S and Nienhuis, A (1985). Developmental regulation of human globin genes. Ann Rev Biochem 54:1071-1108.

Kollias G, Hurst J, deBoer E, Grosveld F (1987). The human β-globin gene contains a downstream developmental specific enhancer. Nucleic Acids Res 15:5739-5747.

Mantovani R, Malgaretta N, Giglioni B, Comi P, Cappellini N, Nicolis S, Ottolenghi S (1987). A protein factor binding to an octamer motif in the the γ-globin promoter disappears upon induction of differentiation and hemoglobin synthesis in K562 cells. Nucleic Acids Res 15:9349-9364.

Trudel M, Magram J, Bruckner L, Constantini F (1987). Upstream ^{G}γ-globin and downstream β-globin sequences required for stage-specific expression in transgenic mice. Mol Cell Biol 7:4024-4029.

Tuan D, Solomon W, Li Q, London I (1985). The β-like-globin gene domain in human erythroid cells. Proc Nat Acad Sci USA 82:6384-6388.

Wall L, DeBoer E, Grosveld F (1988). The human β-globin gene 3' enhancer contains multiple binding sites for an erythroid-specific protein. Genes and Dev 2:1089-1100.

Wright S, Rosenthal A, Flavell R, Grosveld F (1987). DNA sequences required for regulated expression of β-globin genes in murine erythroleukemia cells. Cell 38:265-273.

Zhu, J-d, Allan, M, Paul, J (1984). The chromatin structure of the human ε globin gene: Nuclease hypersensitive sites correlate with multiple initiaiton sites of transcription. Nuc Acid Res 12:9191-9204.

Hemoglobin Switching, Part A: Transcriptional Regulation, pages 193–202
© 1989 Alan R. Liss, Inc.

PROTEINS BINDING TO REGULATORY ELEMENTS 5' TO THE HUMAN β-GLOBIN GENE

Patricia E. Berg,[+] Donna M. Williams,[+] Ruo-Lan Qian,[+] Roger B. Cohen,[*] Moshe Mittelman,[+] and Alan N. Schechter[+]

[+]Laboratory of Chemical Biology, National Institute of Diabetes, and Digestive and Kidney Diseases and [*]National Heart, Lung and Blood Institute, National Institutes of Health Bethesda, Maryland 20892

INTRODUCTION

Although the human β-globin gene has been cloned and sequenced (Poncz et al., 1983), relatively little is known about its regulation. There are promoter sequences located immediately upstream of the cap site, including an ATA box, a CCAAT box and CACCC sequences. Two enhancer sequences have been identified, one intragenic and the other 3' to the gene (Behringer et al., 1987; Kollias et al., 1987; Trudel et al., 1987; Antoniou et al., 1988), as well as a sequence at -160 bp relative to the cap which is necessary for induction of the β-globin gene in mouse erythroleukemia cells (Antoniou et al., 1988). Thus far, however, there is a paucity of information on regulatory sequences further upstream in the 5' flanking sequence of the human gene.

Using DNA deletion analysis and a transient assay expression system, we have identified three regulatory sequences 5' to the human β-globin gene. These include two negative control regions which we call NCR1 and NCR2, located between -610 and -490, and between -338 and -233, respectively. A positive control region, designated PCR, is located between -233 and -185 (Qian et al., 1987; Berg et al., in preparation). NCR1 and NCR2 act as classical silencer elements, since they decrease expression of a

heterologous gene in a position and orientation
independent manner.

In order to understand how these silencer regions
function, we have begun to look for proteins which
specifically bind to them. In this paper we will present
data based on gel electrophoresis mobility shift and DNase
I footprint assays which have allowed us to identify
several proteins from K562 nuclear extracts which bind to
these silencer regions. We have used competition assays
to show that one of these proteins appears to bind equally
well to either silencer element.

RESULTS

We used the gel mobility shift assay (Strauss and
Varshavsky, 1984) to detect proteins capable of binding to
the two silencer elements. Nuclear extracts were prepared
according to Dignam et al. (1983). We analyzed the second

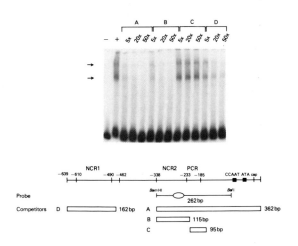

Figure 1. Competition for protein binding to NCR2 DNA. A
schematic diagram of 5' β-globin flanking DNA is shown
below the mobility shift assay. Nucleotide numbers,
relative to the cap at +1, are given for the deletions
used in defining NCR1, NCR2 and PCR. Competitor DNAs are
shown by open boxes, with their positions relative to the
β-globin gene.

silencer with a labelled 262 bp probe (-338 to -76). When
this fragment was incubated with nuclear extracts prepared
from K562 cells, two bands of decreased mobility were
seen, indicated by the arrows in Figure 1. The
specificity of these bands is shown by the fact that the
shifted bands are competed by an unlabelled DNA fragment
which includes the probe (competitor A). A competitor
which encompasses only the second silencer, B, also
competes for both shift bands, while competitor C,
covering only the positive control region, does not.
Thus, the protein causing the shift bands binds
specifically to NCR2. To determine whether this protein
binds to other 5' DNA sequences, competitor D was used.

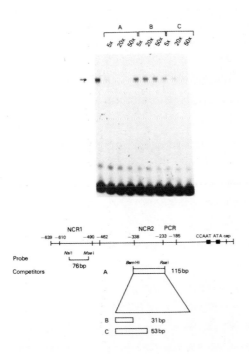

Figure 2. Competition for protein binding to NCR1 DNA by
NCR2 DNA sequences. The 5' end-labelled probe extended
from -576 to -500 bp. The locations of competitors B and
C are shown relative to competitor A.

This DNA covers the first silencer and competes for both
shift bands. These data suggest that a common protein may
bind to both negative regulatory elements.

   To analyze the first silencer region, we performed
mobility shift and DNAse I footprint analyses. Whan a 76
bp probe from NCR1 was incubated with K562 nuclear
extract, a shift band was observed (Fig. 2). Two shift
bands could be produced by increasing the amount of
nuclear extract (data not shown). We wanted to determine
whether there would be cross-competition with NCR2 DNA
fragments for this shift band. Competitor A, of 115 bp,
competed for the shift band, even at a five fold molar
excess. Competitor B, a 31 bp subfragment of A, did not
compete, while C, extending from -338 to -285, did
compete. Therefore, competition occurs for binding of a
protein between both silencers using a probe from either
NCR1 or NCR2. The region binding this protein in NCR2 is
between -338 and -285.

   In order to determine the sequences on NCR1
recognized by this protein, we performed in vitro DNase I
footprint experiments using a 162 bp probe from NCR1 (-639
to -477), 5' end-labeled at -639 (coding strand). As
shown in Fig. 3A, two protected regions, labeled 1 and 2,
are evident. Only a single region of protection
corresponding to region 1 was found on the non-coding
strand, presumably because there are no DNAse I cleavages
over region 2 even in naked DNA (Fig. 3B). The protected
nucleotides are shown in Fig. 3C. Both protected
sequences are located within a region containing a stretch
of 52 alternating purines and pyrimidines. We do not yet
know whether the two regions of protection represent the
effects of a single protein or multiple proteins.

   We wished to determine whether the mobility shift
bands specific for NCR1 were related to the protein(s)
responsible for the DNAse I footprints. When the same DNA
probe used in Fig. 2 was used as a probe in this assay,
two shift bands with the same R_f as the two bands in Fig.
1 were seen (Fig. 4). Three competitors were tested.
Competitor A, which is larger than the probe, showed
specific competition as expected. B, a 29 bp
oligonucleotide which does not include the footprinting
DNA (shown as an ellipse), does not compete. A larger
oligonucleotide, C, which does include the footprinted DNA

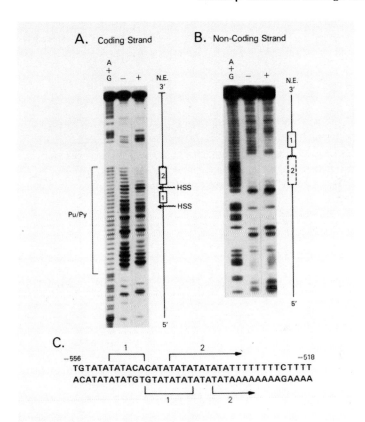

Figure 3. DNAse I protection of NCR1. A. Analysis of the coding strand. The probe was a 162 bp fragment extending from -639 to -477 and was 5' end labelled. The presence or absence of nuclear extract (N.E.) is indicated by a + or -, respectively. Boxes indicate protected regions, HSS denotes DNAse I hypersensitive sites, and Pu/Py shows 52 alternating purines and pyrimidines. B. Analysis of the non-coding strand. The 5' end labelled probe was 204 bp, from -577 to -373. The dotted box indicates where the protected region would be which corresponds to the protected sequence on the coding strand. C. The protected DNA sequence. Protected bases are shown by a solid line. Arrows indicate the regions where it is not possible to determine the end of the footprint.

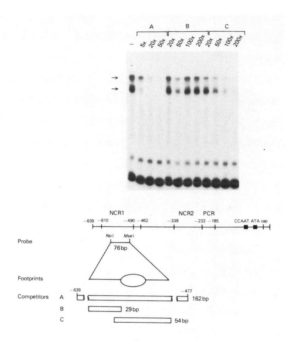

Figure 4. Competition for the footprinting protein of NCR1. The footprinting region is shown as an ellipse on the 76 bp probe, and the positions of the competitors are shown relative to the probe.

does compete. A smaller, 33 bp oligonucleotide encompassing the footprint also shows competition (data not shown). These data suggest that the protein(s) responsible for the DNAse I footprints also give the shift bands seen in Fig. 4.

Further competition experiments were performed using a probe from NCR2 (-338 to -266) and the 33 bp oligonucleotide spanning the footprint in NCR1 as competitor. This DNA competed for the two shift bands (data not shown), from which we conclude that the region from -338 to -285 in NCR2 (competitor C in Fig. 4) and -556 to -523 in NCR1 (the oligonucleotide covering the footprint in NCR1) are capable of interacting with a common protein or proteins.

In vitro DNase I footprint experiments with unfractionated extracts have not yet revealed a footprint over NCR2 in the region from -338 to -285, although this region clearly binds a protein in the mobility shift assay. DNase I footprint assays of NCR2 nevertheless do reveal an obvious footprint which extends from -274 to -256 on the coding strand and -257 to -272 on the non-coding strand (data not shown). Thus, there is at least one additional protein which can bind to human beta-globin 5'-flanking DNA in this region.

DISCUSSION

In this report, we have identified at least two proteins which can bind to two silencer elements located 5' to the human β-globin gene. One of these proteins binds within the first silencer element, NCR1, and gives rise to a DNAse I footprint located between 550 and -527 relative to the cap site (Fig. 3). We have also been able to detect this protein using the mobility shift assay. Competition experiments show that using a probe from NCR1 the two specific mobility shift bands could be competed not only by unlabeled NCR1 DNA but also by unlabeled NCR2 DNA fragments (Figs. 2 and 4).

Two specific mobility shift bands were also observed using a probe from the NCR2 region (Fig. 1). These complexes were also competed by unlabeled NCR2 DNA and unlabeled NCR1 DNA. By using smaller competitor fragments from both NCR1 and NCR2 in cross-competition experiments, we were able to narrow the region responsible for the cross-competition to -556 to -523 in NCR1 and -338 to -285 in NCR2. Although we have determined the boundaries of the NCR1 protein binding domain by DNAse footprint analysis, we have defined the NCR2 binding domain only by mobility shift competition assays. DNAse I footprint analysis of NCR2 DNA using unfractionated K562 nuclear extracts has yet to reveal a footprint over the NCR2 region responsible for cross-competition. Instead, we have detected a footprint outside this region which probably represents the independent binding of another protein to NCR2.

The fact that the same protein(s) binds to both silencer regions leads us to speculate that it may be a repressor. Repressors are well known in bacterial systems, such as the lac operon, the trp operon, and bacteriophage λ (Ptashne, 1986; Lewin, 1987). Repressors have also been found in yeast (Brent, 1985), and have been inferred for mammalian genes such as β-interferon (Goodbourne and Maniatis, 1988). It now appears that the human γ-globin gene may be regulated in part by a repressor protein binding to an octamer sequence 5' to that gene (Mantovani et al., 1987). The silencers we have identified 5' to the β-globin gene may act by binding of a repressor, which acts at a distance to decrease transcription.

Consistent with our repressor hypothesis is the observation that the protein binding to both silencers is also present in hemin-induced K562 cells, as well as HeLa cells (data not shown). Hemin-induced K562 cells are of particular interest since they express embryonic and fetal globin genes but never express β-globin (Rutherford et al., 1979; Alter and Goff, 1980; Benz et al., 1980; Charney and Maniatis, 1983; Dean et al., 1983). This could be due to over-production of a repressor, decreasedet synthesis of an activator, or both. The detection of the NCR1/NCR2 binding protein(s) in induced K562 cells is consistent with its being a repressor which is not turned off after induction. The presence of this protein in HeLa cells, where the β-globin gene is not expressed, is also consistent with its being a repressor protein.

In conclusion, we have shown evidence that at least two proteins bind to sequences located 5' to the human β-globin gene. Our findings suggest that the same protein(s) may bind to two different sites. These sites are within two regions which both have silencer activity, suggesting this protein(s) may act as a repressor.

REFERENCES

Alter BP, Goff SC (1980). Electrophoretic separation of human embryonic globin demonstrates "α-thalassemia" in human leukemic cell line K562. Biochem Biophys Res Commun 94:843-848.

Antoniou M, deBoer E, Habets G, Grosveld F (1988). The human β-globin gene contains multiple regulatory regions: identification of one promoter and two downstream enhancers. EMBO J 7:377-384.

Behringer R, Hammer R, Brinster R, Palmiter R, Townes T (1987). Two 3' sequences direct adult erythroid-specific expression of human β-globin genes in transgenic mice. Proc Natl Acad Sci USA 84:7056-7060.

Benz EJ, Murnane MJ, Tankanow DL, Beoman BW, Mazur EM, Cavallesco C, Jenko T, Snyder EL, Forget BG, Hoffman R (1980). Embryonic-fetal erythroid characteristics of a human leukemic cell line. Proc Natl Acad Sci USA 77:3509-3513.

Brent R (1985). Repression of transcription in yeast. Cell 42:3-4.

Charnay P, Maniatis T (1983). Transcriptional regulation of globin gene expression in the human erythroid cell line K562. Science 220:1281-1283.

Dean A, Ley TJ, Humphries RK, Fordis M, Schechter AN (1983). Inducible transcription of five globin genes in K562 human leukemia cells. Proc Natl Acad Sci USA 80:5515-5519.

Dignam JD, Lebovitz RM, Roeder RG (1983). Accurate transcription initiation by RNA polymerase II in a soluble extract from isolated mammalian nuclei. Nucleic Acids Res 11:1475-1489.

Goodbourn S, Maniatis T (1988). Overlapping positive and negative regulatory domains of the human β-interferon gene. Proc Natl Acad Sci USA 85:1447-1451.

Kollias G, Hurst J, deBoer E, Grosveld F (1987). The human β-globin gene contains a downstream developmental-specific enhancer. Nucleic Acids Res 15:5739-5747.

Lewin B, ed. (1987). Genes III, John Wiley and Sons, p. 223,234-235.

Mantovani R, Malganetti N, Giglioni B, Comi P, Cappellini N, Nicolis S, Ottolenghi S (1987). A protein factor binding to an octamer motif in the γ-globin promoter disappears upon induction of differentiation and hemoglobin synthesis in K562 cells. Nucleic Acids Res 15:9349-9393.

Poncz M, Schwartz E, Ballantine M, Surrey S (1983). Nucleotide sequence analysis of the δβ-globin gene region in humans. J Biol Chem 258:11599-11609.

Ptashne M (1986). Gene regulation by proteins acting
      nearby and at a distance. Nature 322:697-701.
Qian R-L, Williams DM, Cao S-X, Schechter AN, Berg PE
      (1987). Modulation of the β-globin gene promoter by
      5' negative and positive control regions. Blood 70
      Suppl 1:79.
Rutherford TR, Clegg JB, Weatherall DJ (1979). K562 human
      leukaemic cells synthesize embryonic hemoglobin in
      response to haemin. Nature 280:164-165.
Strauss F, Varshavsky A (1984). A protein binds to a
      satellite DNA repeat at three specific sites that
      would be brought into mutual proximity by DNA folding
      in the nucleosome. Cell 37:889-901.
Trudel M, Magram J, Bruckner L, Constantini F (1987).
      Upstream ^{G}γ-globin and downstream β-globin sequences
      required for stage-specific expression in transgenic
      mice. Mol Cell Biol 7:4024-4029.

Hemoglobin Switching, Part A: Transcriptional Regulation, pages 203–215
© 1989 Alan R. Liss, Inc.

ANALYSIS OF HUMAN GAMMA AND BETA GLOBIN GENE
REGULATION USING TRANSGENIC MICE

Carlos Perez-Stable, Jeanne Magram,
Karen Niederreither and Frank Costantini

Department of Genetics and Development, College
of Physicians and Surgeons, Columbia
University, New York, N.Y. 10032

INTRODUCTION

The transgenic mouse has been proven a valuable
tool for the study of cis-acting DNA sequences
involved in the regulation of the human beta-
globin gene family. Both beta- and gamma-globin
genes are expressed tissue-specifically, and with
appropriate developmental stage specificity, when
introduced into the mouse genome (Chada et al.,
1985, 1986; Townes et al., 1985; Kollias et al.,
1986). A variety of transgenic studies, in our
laboratory and elsewhere, have contributed to our
growing knowledge of the sequence elements
involved in globin gene regulation (Trudel et al.,
1987a, 1987b; Behringer et al., 1987; Kollias et
al., 1987; Grosveld et al., 1987). In this paper,
we review some recent findings regarding the
regulation of the beta- and gamma-globin genes.
We first describe experiments aimed at defining
the nature and location of regulatory elements in
the immediate upstream region of the gamma-globin
gene. We then describe a study in which we have
demonstrated that beta-globin regulatory sequences
can be used to target the expression of heterologous genes in a tissue-specific and stage-specific
pattern.

RESULTS AND DISCUSSION

Expression of Human [G]Gamma Globin Genes with
Truncated 5' Flanking Sequences.

    Previous studies have shown that the human
[G]gamma-globin gene, including 1.65 kb of 5'
flanking DNA, is expressed in the embryonic but
not fetal or adult erythroid cells of transgenic
mice (Chada et al., 1986). A variety of data from
other studies has implicated sequences in the
immediate 5' flanking regions of the gamma globin
genes in the process of hemoglobin switching. To
define the 5' sequences required for the activity
of the [G]gamma gene in mouse embryonic erythroid
cells, and for its inactivation in fetal and adult
cells, we tested DNA fragments containing the
[G]gamma gene together with 383, 201 or 136 nt of 5'
flanking DNA. Each fragment was microinjected
into fertilized mouse eggs, transgenic embryos
were identified at 11.5 days of gestation, blood
RNA was isolated and the levels of gamma-globin
mRNA were measured (Figure 1 and Table 1).

    The expression levels of the -383 gamma-globin
transgene were not significantly different from
that of the -1654 transgene, indicating that no
major positive regulatory elements are located
upstream of -383. The -201 transgene was expressed in somewhat fewer transgenic embryos, but the
levels of expression were similar to the -1654
transgene. In contrast, the -136 transgene was
essentially silent. Thus, the major positive cis-
acting elements for expression in mouse embryonic
blood cells appear to be located between -201 and
-136. This region contains a number of sequence
motifs known to bind several nuclear factors, and
believed to be important for the expression of the
gamma-globin genes in human fetal erythroid cells
(Mantovani et al., 1987,1988; Gumucio et al.,
1988; see other papers in this volume). These
include the octamer element, two sites that bind
the erythroid specific factor NFE1, and the "CACCC
box". Our results confirm, in an in vivo developmental system, that one or more of these elements
are essential for gamma-globin gene expression.

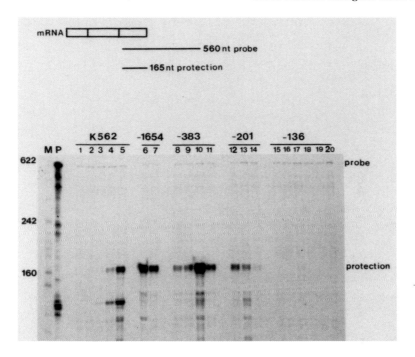

Figure 1. Expression of 5′ truncated ^Ggamma globin genes, measured by RNase protection. Each lane shows the analysis of 2 ug blood RNA from selected 11.5 day embryos. Lanes 1-5 contain 0, .03, .15, .75 and 3 ug of total RNA from induced K562 cells.

The -136 to -201 region may also contain negative regulatory elements involved in hemo-globin switching, as suggested by mutations in this region that are associated with hereditary persistence of fetal hemoglobin (HPFH) in man. Both the -383 and -201 transgenes are appropriat-ely inactive in mouse fetal livers (C.P.-S. and M. Trudel, unpublished data). Therefore, if the inactivity of the gamma globin transgene in mouse fetal liver involves a negative regulatory mechanism, this mechanism does not require sequences upstream of -201. Whether transgenes lacking some of the elements between -201 and -136 may be active in mouse fetal liver, which would support such a model, is currently being examined.

Table 1. Expression of 5' truncated ^{G}gamma-globin transgenes in 11.5 day embryonic blood cells.

Transgene	mRNA Levels (pg/ug total RNA)
-1645[a]	0, 0, 0, 4, 8, 40
-383	0, 0, 0, 2, 4, 10, 400
-201	0, 0, 0, 0, 0, 0, 0, 4, 10, 15
-136	0, 0, 0, 0, 0, 0.4

Each value represents an independent transgenic embryo. The level of endogenous mouse beta-h1 globin mRNA in 11.5 day mouse embryonic blood is 200 pg/ug. a) data from Chada et al., 1986.

Activation of a Beta Globin Gene in Embryonic Blood Cells by Gamma Globin 5' Flanking Sequences.

   We have previously shown that a hybrid gene containing ^{G}gamma globin 5' sequences from -1.65 kb to the initiation codon, fused to the beta-globin structural gene and 3' flanking sequences, was expressed in transgenic mouse embryonic blood cells (Trudel et al., 1987a). In contrast, the beta-globin gene itself is not expressed in the embryonic erythroid cells of transgenic mice (Magram et al., 1985; Townes et al., 1985). Thus, gamma-globin upstream elements are sufficient for expression of hybrid gamma/beta globin genes at the embryonic stage. To better define the sequences responsible for this effect, we tested the expression of a construct containing the beta-globin gene with 127 bp of 5' flanking sequence, fused to ^{G}gamma sequences from -383 to -136 (GB-SX; Figure 2). While the -127 beta-globin gene in isolation (B-CX) was not expressed at significant levels in embryonic blood cells, the addition of the ^{G}gamma globin upstream sequences resulted in high levels of expression (Figure 3 and Table 2). This agrees with the results of Lin et al. (1987), who obtained similar results using essentially the same construct in K562 cells.

^{G}gamma globin sequences from -201 to -136 were also sufficient to activate the -127 beta-globin gene (construct GB-AX), confirming that the -201 to -136 region contains positive regulatory elements for expression in mouse embryonic erythroid cells. However, this construct was expressed at considerably lower levels than

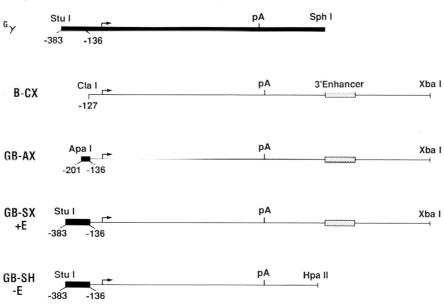

Figure 2. Hybrid gamma/beta globin genes. Gamma-globin sequences are shown by thick lines and beta-globin sequences by thin lines.

Table 2. Expression levels of hybrid gamma/beta globin transgenes in 11.5 day embryonic blood.

Transgene	mRNA Levels (pg/ug total RNA)
B-CX	0, 0, 0, 0, 0, 0.3, 0.4
GB-AX	0, 0, 1, 2, 4, 4, 5, 25
GB-SX	0, 0, 4, 10, 10, 200, 200, 200, 300, 2000
GB-SH	0, 0, 0, 0, 0.5, 1.5, 1.5, 15, 20, 30, 50

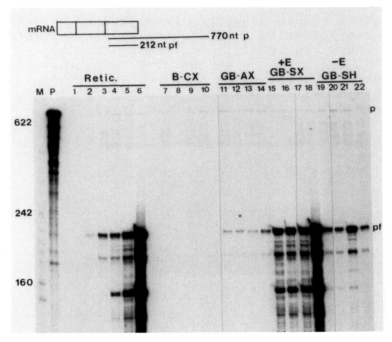

Figure 3. Expression of hybrid gamma/beta globin genes measured by RNase protection. Lanes 1-6 contain 0, .23, 1.1, 5.7, 23 and 230 ng of human reticulocyte total RNA. Lanes 7-22 contain 2ug of total blood RNA from selected transgenic embryos.

construct GB-SX, suggesting the presence of additional positive elements in the -383 to -201 region. Whether the -383 to -201 fragment can function independently to activate the beta-globin gene is currently being investigated.

The Beta-Globin 3' Enhancer Stimulates a Gamma/ Beta Hybrid Gene in Mouse Embryonic Blood Cells.

Several studies have demonstrated that the 3' flanking region of the human beta-globin gene contains an enhancer that functions in mouse fetal and adult erythroid cells (Behringer et al., 1987; Kollias et al., 1987; Trudel et al., 1987b). Although the 3' enhancer is not absolutely required for expression of the beta-globin gene, possibly because the gene also contains an intragenic enhancer (Behringer et al., 1987; Brinster et al., 1988; R. Gelinas, personal communication), deletion of the 3' enhancer results in a 10-fold reduction in the level of expression in mouse fetal liver (Trudel et al., 1987b). Because the beta-globin gene itself is not expressed in mouse 11.5 day embryonic erythroid cells, it seemed unlikely that the 3' enhancer would be active at this stage of development. Surprisingly, we found that the expression of the gamma/beta hybrid construct in mouse embryonic blood cells was reduced approximately 10-fold by deletion of the beta-globin 3' enhancer (construct GB-SH). Thus, the 3' enhancer appears to stimulate transcription in embryonic as well as fetal and adult mouse erythroid cells. As discussed below, this suggests that the stage-specific expression of the beta-globin gene may be controlled by some other cis-acting regulatory element.

Erythroid-Specific Expression of a Heterologous Gene under the Control of Beta Globin Regulatory Elements.

The beta-globin enhancers were first defined by their ability to activate a human gamma globin gene in fetal and adult erythroid cells of transgenic mice (Behringer et al., 1987; Kollias et al., 1987; Trudel et al., 1987b), as well as by the effects of deleting sequences from the beta-globin gene itself (Trudel et al., 1987b; Brinster et al., 1988; R. Gelinas, personal communication). Sequences with enhancer activity include a region approximately 600-900 bp 3' to the poly (A) site, as well as a less extensively characterized region in the second intron and/or third exon. We are interested in the possibility that the beta-globin enhancers may be useful to target the expression of heterologous genes of interest to erythroid cells. We have therefore examined the ability of the two beta-globin enhancers, in combination, to direct tissue-specific and stage-specific transcription from a heterologous promoter in transgenic mice.

The beta globin DNA fragment we tested was a BamHI-XbaI fragment, including the second intron, the third exon and 1.5 kb of 3' flanking DNA. This was joined to a SphI-BamHI fragment from the SV40 early region, which contained the early promoter but lacked a complete copy of the SV40 enhancer, and was thus inactive in the absence of an added enhancer. The SV40 sequences also contained a deletion in the sequence encoding large T-antigen, to eliminate the transforming effect of T-antigen. In one experiment transgenic fetuses were identified at day 16 of gestation, total RNA was isolated from several tissues, and SV40 transcripts were detected by RNase protection analysis (Figure 4 and data not shown). In six transgenic fetuses carrying a control SV40 construct (tSV40) lacking beta-globin sequences, no expression was observed in any of the tissues examined. In contrast, two transgenic fetuses carrying the beta-globin SV40 construct (B3'tSV40) expressed the SV40 RNA in the liver, the major fetal erythroid tissue, but not in the four non-erythroid tissues examined. As shown in Figure 4 for one such fetus, B26, liver RNA protects a series of fragments approximately 150 nt in length, consistent with the expected cap sites of the SV40 early promoter. The other five trans-genic fetuses did not contain detectable SV40 transcripts in any of the five tissues examined.

To confirm these results in adult mice, we injected another series of eggs with the B3'tSV40 construct and established four transgenic lines. In two lines, 20 and 30, SV40 transcripts were observed in peripheral blood, bone marrow and anemic spleen, but were absent or seen at much lower levels (possibly due to blood contamination) in non-erythroid tissues (data not shown). No expression was observed in the other two lines.

These experiments show that a beta-globin fragment containing the second intron, third exon and 3' flanking sequences is capable of enhancing transcription from a heterologous promoter in a tissue-specific pattern. In each of four indepen-dent transgenic lines or fetuses in which synthesis of SV40 early RNA was detected, it was either specific to erythroid tissue or highest in erythroid tissue. The level of expression of this hybrid construct cannot be compared in a meaningful way to that of the normal beta-globin gene, because the relative stabilities of beta-globin and SV40 mRNAs in different mouse tissues are not known. Nevertheless, the pattern of expression of the hybrid construct indicates that the beta-globin enhancers are tissue-specific in their activity. As the BamHI-XbaI DNA fragment appears to contain two distinct regions with enhancer activity, we cannot determine from the present experiments whether one or both of these elements

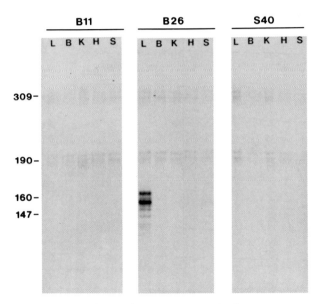

Figure 4. Erythroid tissue-specific expression of the SV40 early region under the control of beta-globin regulatory elements. RNA from liver (L), brain (B), kidney (K), heart (H) and skin (S) of three 16 day transgenic fetuses was analyzed by RNAse protection with an SV40 early region 5' probe. B11 and B26 carry the B3'tSV40 construct, S40 carries the tSV40 control construct. Ten ug of total RNA from liver, brain and skin, or two ug of total RNA from kidney and heart, were used for each reaction.

is responsible for the tissue-specific expression observed. Nevertheless, our results indicate that the BamHI-XbaI fragment could be useful to direct the expression of other heterologous gene products to erythroid cells in transgenic mice.

Beta Globin Enhancers Activate an SV40 Promoter in a Developmental Stage-Specific Pattern.

To test whether B3'tSV40 construct was subject to developmental regulation, we analyzed total RNA from blood cells of 11.5 day transgenic embryos, and livers of 16 day transgenic fetuses in lines 20 and 30. As shown in Figure 5, SV40 early RNA was detected in the fetal livers as well as in adult anemic spleens, but not in embryonic blood from either transgenic line. As a control for the integrity of the embryonic blood cell RNA, the same RNAs were hybridized with a probe specific to the endogenous embryonic Bh1 globin gene. Both embryonic blood RNA samples hybridized to the Bh1 probe, whereas the fetal livers and adult spleens were negative for Bh1 mRNA (data not shown). Thus, the beta-globin intragenic and 3' enhancers, as a pair, appear to be active in mouse fetal and adult but not embryonic erythroid cells.

This result may appear inconsistent with the observation that the 3' enhancer stimulates, by approximately 10-fold, the expression of a gamma/beta hybrid gene in mouse embryonic blood cells (construct GB-SX vs. GB-SH, Figures 2 and 3). However, the two experiments are not strictly comparable because the B3'tSV40 construct includes beta-globin intragenic sequences as well as the 3' enhancer. One possibility that could reconcile the two observations is that the intragenic beta-globin regulatory element may be specific for fetal and adult stage erythroid cells, and may be dominant over the 3' enhancer when both are present. Another difference between the two experiments is that the enhancer(s) must interact in one case with the SV40 promoter, and in the other case with a beta-globin promoter and/or gamma-globin upstream sequences. As protein-protein interaction between factors bound to the

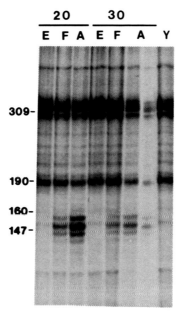

FIG 5. Developmental stage-specific expression of the beta-globin/SV40 fusion gene in transgenic lines 20 and 30. Total RNA was isolated from blood cells of transgenic embryos at day 11 of gestation (E), livers of day 16 transgenic fetuses (F), and adult transgenic mouse spleens after treatment with phenylhydrazine. Y, yeast RNA. A. Ten ug of each RNA was analyzed by RNase protection using an SV40 probe (sample 30A was inadvertently divided between two lanes.) The faint band at 150 nt seen in the embryonic RNA samples is also seen in the yeast RNA sample, and is therefore a background band.

promoter and enhancer are thought to mediate enhancer function, and different factors presumably bind to the SV40 and beta-globin promoters in embryonic erythroid cells, it is not surprising that differences in enhancer activity might be observed. Clearly, an understanding of the mechanisms governing the stage-specific expression of the beta-globin gene is going to require a detailed characterization of the many trans-acting factors involved.

REFERENCES

Behringer RR, Hammer RE, Brinster RL, Palmiter RD, Townes TM (1987). Two 3′ sequences direct adult erythroid specific expression of human beta-globin genes in transgenic mice. Proc Nat Acad Sci USA 84:7056-7060.

Brinster RL, Allen JM, Behringer RR, Gelinas RE, Palmiter RD (1988). Introns increase transcriptional efficiency in transgenic mice. Proc Nat Acad Sci USA 85:836-840.

Chada K, Magram J, Raphael K, Radice G, Lacy E, Costantini F (1985). Specific expression of a foreign beta globin gene in erythroid cells of transgenic mice. Nature 314:377-380.

Chada K, Magram J, Costantini F (1986). An embryonic pattern of expression of a human fetal globin gene in transgenic mice. Nature 319: 685-688.

Grosveld F, van Assendelft GB, Greaves DR, Kollias G (1987). Position-independent, high-level expression of the human beta-globin gene in transgenic mice. Cell 51:975-985.

Gumucio D, Rood K, Gray T, Riordan M, Sartor C, Collins F (1988). Nuclear proteins that bind the human gamma globin promoter: alterations in binding produced by point mutations associated with hereditary persistence of fetal hemoglobin. Molec Cell Biol 8:5310-5322.

Kollias G, Wrighton N, Hurst J, Grosveld F (1986). Regulated expression of human [A]gamma-, beta-, and hybrid gamma-beta globin genes in transgenic mice: manipulation of the developmental expression patterns. Cell 46:89-94.

Kollias G, Hurst J, DeBoer E, Grosveld, F (1987). A tissue and developmental specific enhancer is located downstream from the human beta-globin gene. Nucl Acids Res 15:5739-5747.

Lin H, Anagnou N, Rutherford T, Shimada T, Nienhuis, A (1987). Activation of the human beta globin promoter in K562 cells by DNA sequences 5' to the fetal gamma or embryonic zeta globin genes. J Clin Invest 80:374-380.

Magram J, Chada K, Costantini F (1985). Developmental regulation of a cloned adult beta globin gene in transgenic mice. Nature 315:338-340.

Mantovani R, Malgaretti N, Ciglioni B, Comi P, Cappellini N, Nicolis S, Ottolenghi S (1987). A protein factor binding to an octamer motif in the gamma globin promoter disappears upon induction of differentiation and hemoglobin synthesis in K562 cells. Nucl Acids Res 15:9349-9365.

Mantovani R, Malgaretti N, Nicolis S, Ronchi A, Giglioni B, Ottolenghi S (1988). The effects of HPFH mutations in the human gamma globin promoter on binding of ubiquitous and erythroid specific nuclear factors. Nucl Acids Res 16:7783-7797.

Townes TM, Lingrel JB, Chen HY, Brinster RL, Palmiter RD (1985). Erythroid-specific expression of human beta-globin genes in transgenic mice. EMBO J 4:1715-1723.

Trudel M, Magram J, Bruckner L, Costantini F (1987a). Upstream ^{G}gamma globin and downstream beta globin sequences required for stage-specific expression in transgenic mice. Molec Cell Biol 7:4024-4029.

Trudel M, Costantini F (1987b). A 3' enhancer contributes to the stage-specific expression of the human beta globin gene. Genes and Development 1:954-961.

Hemoglobin Switching, Part A: Transcriptional Regulation, pages 217–228
© 1989 Alan R. Liss, Inc.

AN ERYTHROID-SPECIFIC DNA BINDING FACTOR MEDIATES INCREASED γ-GLOBIN EXPRESSION IN HEREDITARY PERSISTENCE OF FETAL HEMOGLOBIN (HPFH)

David I.K. Martin and Stuart H. Orkin

Department of Hematology/Oncology, Children's Hospital and Howard Hughes Medical Institute, Boston, Massachusetts 02115

INTRODUCTION

The mechanisms controlling the complex tissue-specific and developmental regulation of the β-like globin genes and those accounting for increased γ-globin expression seen in hereditary persistence of fetal hemoglobin (HPFH) syndromes are incompletely understood (Stamatoyannopoulos and Nienhuis 1987). Three lines of evidence suggest that cis-regulatory sequences controlling expression of γ-globin genes reside within their upstream sequences. First, introduction of hybrid constructs of the γ-promoter from -385 to +34 with the neomycin-resistance gene allowed expression of drug selection in human erythroleukemia K562 cells but not in non-erythroid cells (Rutherford and Nienhuis 1987). Sequences from -257 to -140 conferred transcriptional activity on an otherwise silent β-globin promoter in K562 cells, which suggested the presence of an "activator" element (Lin, Anagnou et al. 1987). Second, the existence of erythroid-specific DNAse I hypersensitivity sites within the γ-globin promoters has suggested that these may interact with sequence-specific nuclear proteins, as described for the chicken globin genes by Felsenfeld and his colleagues (Tuan, Solomon et al. 1985; Kemper, Jackson et al. 1987). Third, single base substitutions in the γ-globin promoters in association with HPFH syndromes are indirect evidence for the presence of critical cis-acting sequences. Mutations at -161, -175, and -202 of the Gγ-

promoter and -198, -196, and -202 of the Aγ-promoter are
linked to higher levels of expression of Gγ- and Aγ-
globin, respectively (Stamatoyannopoulos and Nienhuis
1987). The -175 T-C Gγ-promoter substitution (Surrey,
Delgrosso et al. 1988; Ottolenghi, Nicolis et al. 1988) is
at one end of an octanucleotide motif (ATGCAAAT) found in
several promoter and enhancer elements and known to bind
ubiquitous (NF-A1 or OTF-1) and cell-type specific (NF-A2
or OTF-2) factors (Scheidereit, Heguy et al. 1987; Staudt,
Singh et al. 1986). In view of these findings it appears
likely that the upstream γ-globin region from -257 to -140
contains one or more binding sites for trans-acting
factors.

       In the work summarized here we have sought to define
the factors that bind to the -257 to -140 region of γ-
promoter and to provide evidence for their role(s) in
normal γ-gene expression and the HPFH syndrome. We have
found a novel erythroid-specific factor in K562 cells that
recognizes a conserved motif within sequences from -195 to
-170. Binding of this factor mediates the increased
expression associated with the -175 T-C HPFH mutation.
Furthermore, this factor appears to be the human
counterpart of independently identified factors, Eryf 1
(Evans et al. 1988) and NF-E1 (Wall et al. 1988), that
bind to other cis-acting elements of globin gene promoters
and enhancers in chicken, mouse, and human.

RESULTS

       Identification and Characterization of Nuclear DNA-
Binding Factors: We first sought evidence for nuclear
factors in γ-globin producing K562 cells capable of
binding in vitro to a -257 to -140 γ-promoter fragment.
Incubation with crude nuclear extracts of K562 and non-
erythroid Hela cells produced gel-retardation complexes
(Strauss and Varshovsky 1984; Fried and Crothers 1981)
displayed in Figure 1. Complexes designated 1A and 1B have
been observed only with extracts derived from erythroid
cells (K562, MEL, and HEL). In contrast, complex 2A
results from the binding of the ubiquitous octamer-binding
factor (NF-A1/OTF-1). The different specificities of the
factors were established by competition of gel-retardation
complexes using extract crudely fractionated by heparin-

Figure 1. Gel-retardation of Hinf
I-Nco I γ-promoter (-260 to -140)
fragment incubated with K562 and Hela
cell nuclear extracts.

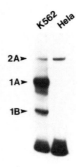

sepharose chromatography. As shown in Figure 2, formation
of complex 2A is specifically competed by -193 to -169
sequences and by the core octamer sequence but not by a
heterologous oligonucleotide or by DNA containing the -175
T-C HPFH substitution. Conversely, formation of the
erythroid-specific complexes 1A and 1B is competed by
homologous sequences or the -175 T-C mutant DNA but not by
the core octamer.

Figure 2. Competition
of gel-retardation
patterns by oligo-
nucleotides. Heparin-
agarose fractionated
K562 cell nuclear
extract was used with
the Hinf I-Nco I
fragment as probe with-
out competitor (-) or
with the addition
of a 400-fold excess
of (1) 26 bp sequence
from -193 to -169; (2)
32 bp sequence

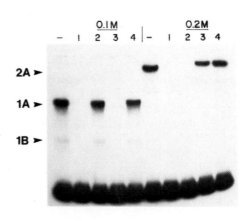

containing repeats of the core octamer (ATGCAAAT); (3) 26
bp identical to (1) except for T-C substitution at -175;
and (4) a heterologous sequence. 0.1M Fraction contains
the erythroid-specific factor (GF-1) and the 0.2M fraction
NF-A1.

DNase I footprinting (Jones, Yamamoto et al. 1985)
and methylation interference (Siebenlist and Gilbert 1980)
assays have been used to characterize these DNA-binding
activities further. As summarized in Figure 3, complex 2A
contacts bases only within the octamer itself. These data
are consistent with its designation as NF-A1/OTF-1.
Fractions with activity yielding complexes 1A and 1B
footprint the promoter from -195 to -170, but contact
bases on either side of the octamer within direct repeats
of five bases (TATCT or AGATA). The similarity of DNA
contact sites in these repeats suggests the existence of
two domains that might bind a single or two protein
molecules. We have found that mutation of contact sites
in both contacted regions is required to abolish binding
of the erythroid factor, designated GF-1, to the -195 to
-170 region.

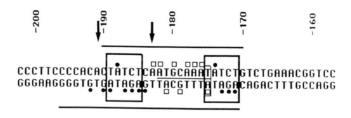

Figure 3. Summary of DNase I footprinting and methylation
interference assays. Horizontal lines denotes the
sequences of each strand protected by the erythroid-
specific factor present in K562 extracts. Open squares
and closed circles indicate methylated bases that
interfere with binding in complex 2A and 1A, respectively.
Downward arrows indicate the positions of DNase
hypersensitive sites.

While our work was in preparation Evans et al
(Evans, Reitman et al. 1988) described an erythroid-
specific globin DNA-binding activity in chicken (Eryf 1)
that recognizes a proposed consensus {(A/T)GATA(A/G)}.
GF-1, and the independently identified human factor NF-E1
(Wall, deBoer et al. 1988; Mantovani, Malgaretti et al.

1988), appears to be the counterpart of this chicken
factor. Specific competition of GF-1 binding to the γ-
promoter by sequences derived from the 3'-β enhancers of
chicken or human origin supports this conclusion (Figure
4).

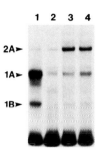

Figure 4. Competition of gel-
retardation complexes (1A and 1B)
by sequence derived from the 3'
β-enhancers of human (lane 3)
and chicken (lane 4) origin.
Lane 2: competition with homologous
human γ-promoter sequences.

Correlation of Promoter Activity with In vitro DNA-
Binding: Having observed binding of two factors in vitro
to overlapping sequences in the -195 to -170 region, we
evaluated their relevance to the enhanced expression seen
in HPFH. Promoter activity was first assessed using a
transient expression assay in which wild-type and mutant
promoters (encompassing sequences from -385 to +50) linked
to the human growth hormone gene (GH) as a reporter
(Selden, Burke-Howie et al. 1986) were introduced into
K562, MEL, and non-erythroid cell lines. Data are
summarized in Table 1.

When introduced into K562 in supercoiled plasmid,
the -175 T-C HPFH promoter directed produced of 4.4 times
as much GH as the wild-type. Expression directed by the
mutant promoter was not increased in two non-erythroid
cell lines (Hela and PLB-985). Since increased γ-globin
expression in HPFH in vivo is manifest in erythoblasts
that synthesize predominantly β-globin, we also assayed
the behavior of the -175 T-C promoter in MEL cells. The
relative enhancement of expression was comparable to that
seen in K562 cells. From these findings we conclude that
the -175 T-C substitution found naturally in association
with HPFH increases promoter activity in a cell-type
specific manner. The phenotype is demonstrable in
erythroid cells of either fetal/embryonic (K562) or adult
(MEL) type.

Table 1

Relative Expression of γ-Promoter/GH and γ-Promoter/Neo Constructs					
	Transient Expression				Stable Selection
Promoter	K562	MEL	HeLa	PLB-985	K562
Supercoiled plasmids					
Wild-type	1.0	1.0	1.0	1.0	1.0
-175 (T→C)	4.4[*]	4.8[*]	1.1	1.2	3.1[**]
-175,-186,-172 (T→C,C→A,C→A)	1.1	0.6			0.6[*]
-186,-172 (C→A,C→A)	1.0				0.5[**]
-180,-179,-178 (GCA→AAG)	0.6				
-179 (C→A)	0.8				
-179,-172 (C→A,C→A)	0.9				
-175,-186 (T→C,C→A)	1.4				
-172 (C→A)	0.9				
Linearized plasmids					
Wild-type	1.0				
-175 (T→C)	1.4				

[*]p<0.001
[**]p<0.01
[*]p<0.05

As the -175 T-C substitution greatly reduces binding of NF-A1/OTF-1 (Figure 2), it has been suggested that alleviation of negative regulation by this factor might account for the HPFH phenotype (Mantovani, Malgaretti et al. 1987; Mantovani, Malgaretti et al. 1987). to address this model directly we introduced base substitutions elsewhere in the core octamer. Expression was not enhanced either by a triple substitution (-180 to -178) or a single base change (-179) in the octamer (Table 1). In vitro binding to these mutant promoters by the octamer-binding factor, but not by GF-1, is drastically reduced in either instance (not shown). We conclude, therefore, that the altered expression of the -175 T-C promoter in erythroid cells is not due per se to reduced binding of a negatively acting octamer-binding factor.

To address the role of GF-1 binding in γ-expression we mutated its DNA contact sites flanking the octamer in the context of the wild-type and -175 T-C promoters. Introduction of C-A substitutions at -186 and -172 or a single C-A change at -186 in the -175 T-C promoter reduced expression in K562 cells to the wild-type level(Table 1). On the basis of these and additional mutations, we infer that enhanced, erythroid-specific expression of the -175 T-C promoter is dependent on binding of GF-1.

Whereas enhanced expression appears to require certain GF-1 contact sites, mutations at -186, -172, or -186/-172 revealed little effect on transient expression in the context of the wild-type promoter (Table 1). Since we have observed that γ-GH constructs exhibit basal promoter activity that is not cell-type specific, we also assessed the role of GF-1 binding in the -195 to -170 region with a stable selection assay which has previously been shown to display tissue-specificity (Rutherford and Nienhuis 1987). γ-Promoter/neo-resistance gene constructs were introduced into K562 cells and the number of G418-resistant colonies was scored. Mutations at -186/-172 in the presence or absence of the -175 T-C base change reproducibly lowered the frequency of G418-resistant clones by about 50% (relative to wild-type). In agreement with the transient expression data for the -175 T-C mutant, the -175 T-C/neo-gene plasmid yielded approximately 3-fold more G418-resistant colonies than the wild-type (Table 1). The modest reduction in G418-selection of K562 cells transfected with constructs containing mutations that abolish GF-1 binding to the -195 to -170 region in vitro suggests that additional regulatory elements in the -385 to +50 promoter fragment also contribute to erythroid specificity. There may also be potential binding sites for GF-1 (or NF-E1) in the CCAAT box region (Wall, deBoer et al. 1988).

Binding of GF-1 to the -175 T-C HPFH Promoter:
Because our data directly implicate GF-1 in mediating the increased promoter activity of the -175 T-C mutation, we have examined the interaction of the mutant promoter and the factor in vitro. Although it has been suggested recently that the affinity of the -175 T-C mutant promoter for presumably the same erythroid-specific factor may be

increased 5-fold (Mantovani, Malgaretti et al. 1988), we
do not find this to be the case. As shown in Figure 5,
with either crude or fractionated nuclear extracts, we
observe no difference in the affinity of GF-1 for the
wild-type and mutant promoters. While the -175
substitution alters the proximal TATCT (AGATA) motif to
CATCT (AGATG), GF-1 can still bind to the proximal domain,
as revealed by methylation interference assay (not shown).
Nonetheless, the DNase I footprint of the mutant DNA shows
decreased protection of the proximal motif (not shown),
implying an alteration in the precise manner in which the
factor recognizes the promoter in this region. Finally,
we have observed that linearization of the -175 T-C
promoter/GH construct prior to transient expression in
K562 cells largely abolishes enhanced expression (Table
1). Taken together, our findings suggest that the
increased expression characteristic of the HPFH mutant is
due to an interaction of GF-1 and the promoter that is
influenced by both primary sequence and DNA secondary
structure.

Figure 5. Competition of
erythroid-specific gel-
retardation complexes by
wild-type and -175 T-C
oligonucleotides. Probe:
Hinf I/Nco I fragment.
Competitor: -193 to -169
sequences of wild-type (WT)
or -175 T-C mutant at 1, 4,
10, 20 mg. Extract used

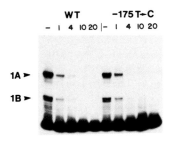

was K562 nuclear proteins fractionated on heparin-
sepharose. Identical results are obtained with
unfractionated extract (not shown).

DISCUSSION

     Through study of proteins capable of <u>in vitro</u>
sequence-specific binding to the γ-globin promoter we have
identified a nuclear factor (designated here (GF-1)
restricted to erythroid cells of either embryonic/fetal or
adult type. As GF-1 appears to recognize the same

sequences bound by factors Eryf 1 and NF-E1, identified
independently in the analysis of promoters and enhancers
of chicken, mouse, and human globin genes (Mantovani,
Malgaretti et al. 1988; Evans, Reitman et al. 1988; Wall,
deBoer et al. 1988), it is likely to play a broad role in
the regulation of gene expression in the erythroid
lineage.

Our systematic mutagenesis of the GF-1 binding site
in the -195 to -170 region supports the contention that
the factor positively regulates γ-globin expression.
Although we cannot preclude a role for the conserved
octamer sequences (-183 to -175) in γ-gene expression,
mutations of the octamer that interfere with binding of
the ubiquitous factor NF-A1 (OTF-1) do not generally lead
to inappropriately high expression, as evident with the
-175 T-C HPFH substitution. We find that increased
expression directed by the -175 T-C HPFH promoter in
erythroid cells is dependent on integrity of the GF-1
binding site(s) in the -195 to -170 region. At present,
the precise effects of the -175 T-C mutation on GF-1
binding and activity on the γ-promoter cannot be defined.
Nonetheless, the evidence summarized above indicates an
altered interaction of the factor with this HPFH promoter,
one that must underlie the augmented expression observed
in an erythroid cell environment. As such, this form of
HPFH is the first inherited disorder that can be ascribed
to the action of a cell-specific DNA-binding factor on a
mutant promoter.

Potential binding sites for GF-1 (Eryf 1, NF-E1) are
widely distributed among the regulatory elements of the
globin genes. Evidence presented above and elsewhere
establishes a functional role for GF-1 binding sites in
the human γ-promoter and in the chicken 3' β-enhancer
(Evans, Reitman et al. 1988). Thus, this factor appears
to act on genes expressed at different developmental
stages and may, in part, serve to determine the fate of
the erythroid lineage. Our work demonstrates that GF-1
can act as a positive regulator of γ-gene expression and
illustrates how a single base change in the sequence to
which it binds may result in substantially increased
promoter activity. Further understanding of how the -175
T-C mutation ultimately affects GF-1 function on the γ-

promoter may suggest how a simple binding motif can participate in complex, regulated gene expression. Purification and structural analysis of GF-1 (and its counterparts) should provide additional insights into the developmental biology of erythroid cells.

REFERENCES

Evans M, Reitman M, Felenfeld G (1988). An erythrocyte-specific DNA-binding factor recognizes a regulatory sequence common to all chicken globin genes. Proc Natl Acad Sci (USA) 85:5976-5980.

Fried M, Crothers DM (1981). Equilibria and kinetics of lac repressor-operator interactions by polyacrylamide gel electrophoresis. Nucl Acids Res 9:6505-6523.

Jones KA, Yamamoto KR, Tjian R (1985). Two distinct transcription factors bind to the HSV thymidine kinase promoter in vitro. Cell 42:559-572.

Kemper B, Jackson PD, Felsenfeld G (1987). Protein-binding sites within the 5' DNase I-hypersensitive region of the chicken αD-globin gene. Mol Cell Biol 7:2059-2069.

Lin HJ, Anagnou NP, Rutherford TR, Shimada T, Nienhuis AW (1987). Activation of the human β-globin promoter in K562 cells by DNA sequences 5' to the fetal or embryonic-globin genes. J Clin Invest 80:374-380.

Mantovani R, Malgaretti N, Giglioni B, Comi P, Cappelini N, Nicolis S, Ottolenghi S (1987). A protein factor binding to an octamer motif in the γ-globin promoter disappears upon induction of differentiation and hemoglobin synthesis in K562 cells. Nucl Acids Res 15:9349-9364.

Mantovani R, Malgaretti N, Nicolis S, Ronchi A, Giglioni B, Ottolenghi S (1988). The effects of HPFH mutations in the human γ-promoter on binding of ubiquitous and erythroid specific nuclear factors. Nucl Acids Res 16:7783-7797.

Ottolenghi S, Nicolis S, Taramelli R, Malgaretti N, Mantovani R, Comi P, Giglioni B, Longinotti M, Dore R, Oggiano L, Pistidda P, Serra, A, Camaschella c, Saglio G (1988). Sardinian Gγ-HPFH: a T to C substitution in a conserved "octamer" sequence in the Gγ-globin promoter. Blood 71:815-817.

Rutherford T, Nienhuis AW (1987). Human globin gene promoter sequences are sufficient for specific expression of a hybrid gene transfected into tissue culture cells. Mol Cell Biol 7:398-402.

Scheidereit C, Heguy A, Roeder RG (1987). Identification and purification of a human lymphoid-specific octamer-binding protein (OTF-2) that activates transcription of an immunoglobulin promoter in vitro. Cell 51:783-793.

Selden RF, Burke-Howie K, Rowe ME, Goodman HM, Moore DD (1986). Human growth hormone as a reporter gene in regulation studies employing transient gene expression. Mol Cell Biol 6:3173-3179.

Siebenlist U, Gilbert W (1980). Contacts between Escherichia coli RNA polymerase and an early promoter of phage T7. Proc Natl Acad Sci (USA) 77:122-126.

Stamatoyannopoulos G, Nienhuis AW (1987). Hemoglobin switching. In Stamatoyannopoulos G, Nienhuis AW, Leder P, Majerus PW, (eds): "The Molecular Basis of Blood Diseases," Philadelphia, PA: W.B. Saunders Company, pp 66-105.

Staudt LM, Singh H, Sen R, Wirth T, Sharp PA, Baltimore D (1986). A lymphoid-specific protein binding to the octamer motif of immunoglobulin genes. Nature 323:640-643.

Strauss F, Varshovsky A (1984). A protein binds to a satellite DNA repeat at three specific sites that would be brought into mutual proximity by DNA folding in the nucleosome. Cell 37:889-901.

Surrey S, Delgrosso K, Malladi P, Schwartz E (1988). A single-base change at position -175 in the 5'-flanking region of the Gγ-gene from a black with Gγ-HPFH. Blood 71:807-810.

Tuan D, Solomon W, Li Q, London IM (1985). The β-like globin gene domain in human erythroid cells. Proc Natl Acad Sci (USA) 82:6384-6388.

Wall L, deBoer E, Grosveld F (1988). The human β-globin gene 3' enhancer contains multiple binding sites for an erythroid-specific protein. Genes and Devel 2:1089-1100.

Hemoglobin Switching, Part A: Transcriptional Regulation, pages 229–236
© 1989 Alan R. Liss, Inc.

ALTERED BINDING TO THE γ-GLOBIN PROMOTER OF TWO ERYTHROID SPECIFIC NUCLEAR PROTEINS IN DIFFERENT HPFH SYNDROMES.

Ottolenghi S., Mantovani R., Nicolis S., Ronchi A., Malgaretti N., [o]Giglioni B. and Gilman J.

Dipartimento di Genetica e di Biologia dei Microrganismi, Università di Milano, [o]Centro per lo Studio della Patologia Cellulare del CNR, Milano, [—]Department of Cell and Molecular Biology, Medical College of Georgia, Augusta, Georgia, U.S.A.

INTRODUCTION

Hereditary Persistence of Fetal Hemoglobin (HPFH) is a benign condition characterized by the continued expression of one or both fetal globin genes ($^{G}\gamma$ and $^{A}\gamma$) during the adult period. In the non-deletion variety of HPFH, either the $^{G}\gamma$ - or the $^{A}\gamma$-globin gene is expressed in the adult period at levels often exceeding by 50-fold or more the expected values; point mutations in the promoter of the overexpressed γ-globin gene have been found consistently, and represent the only detectable difference in the HPFH γ-globin gene relative to the normal gene; moreover, these mutations have never been detected in normal individuals, have arisen independently in different populations, and, when present, are consistently associated with the HPFH phenotype (Collins et al., 1984; Waber et al., 1986; Ottolenghi et al., 1988 a,b; Yang et al., 1988). This genetic evidence strongly indicates a causal relationship between the mutation and the HPFH phenotype. Two hypotheses have been put forward to explain the molecular mechanism of these defects: the mutation might either increase the affinity of the promoter for an activator, or decrease the affinity for an inhibitor protein. Here we discuss the possible role of mutations in two parts (upstream and CCAAT box regions) of the γ-globin promoter and of various nuclear factors interacting with those regions in the abnormal regulation of γ-globin genes in HPFH (see Figure 1).

RESULTS

The upstream promoter region.
Several HPFH mutations (-202, -198, -196, -175, -161, -158) map in the upstream region of the γ-globin promoter (Stamatoyannopoulos and Nienhuis, 1987; Gilman et al., 1988 a; Ottolenghi et al., 1988 b;)in most of these cases

there is little information as to the effect of the mutation on binding of nuclear proteins.

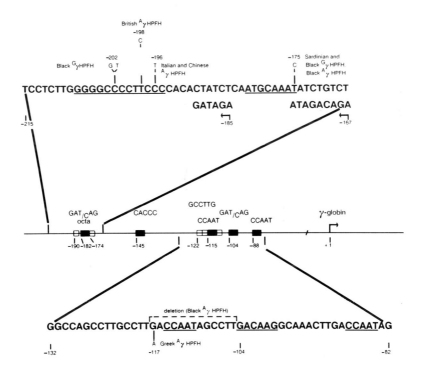

Figure 1. Structure of the human γ-globin promoter and of mutations associated with HPFH.

Recent research demonstrates that the ubiquitous octamer binding factor (OTF$_1$, Fletcher et al., 1987) is able to bind to its recognition sequence (ATGCAAAT) at positions -182 to -175, while the erythroid specific factor NFE-1 binds to two sites immediately upstream and downstream to the octamer, characterized by the consensus GAT$_{/C}$AGT$_{/A}$ (Mantovani et al., 1988). A T→C mutation at position -175 (Ottolenghi et al., 1988 b; Stoming et al., 1988; Surrey et al., 1988) causes $^G\gamma$- or $^A\gamma$-HPFH; we have investigated the effect of this mutation on OTF-1 and NFE-1 binding, and shown that OTF-1 binding is greatly decreased by the mutation, while NFE-1 binding is slightly increased (Mantovani et al., 1988). These results suggest two hypotheses explaining the HPFH phenotype caused by the -175 mutation; OTF-1 might act as a repressor or NFE-1 might act as an activator. To examine these possibilities, we fused the normal and various mutated γ-globin promoters to a reporter gene (CAT, chloramphenicolacetyltransferase) within the plasmid pSVo (Gorman et al., 1982) and transfected the erythroid cell line K562. Figure 2 shows that the mutated -175 HPFH promoter is clearly more active in driving the expression of the CAT gene than its normal counterpart; on the other hand, an A→G mutation at position -177, (which prevents OTF-1 binding while not increasing NFE-1 binding, Mantovani et al., 1988) has no effect on γ-globin promoter activity (data not shown). All CAT plasmids driven by γ-globin promoters have very low activities in non erythroid cells, like HeLa, BJA-B (B-lymphocytes) and HL-60 (granulocyte-monocyte precursor cells); in these cells the -175 mutation has no effect (not shown), indicating that erythroid specific nuclear proteins may be required for conferring the HPFH phenotype. Thus, the present data correlate the increased expression of the -175 promoter to increased binding of NFE-1, not to decreased binding of OTF-1, suggesting that NFE-1 may be an erythroid specific activator of the γ-globin promoter.

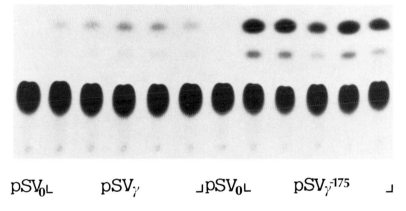

pSV_0L pSV_γ ⌐pSV_0L pSV_γ^{-175} ⌐

Figure 2. Transfection of uninduced K562 cells with γ-globin promoter-driven CAT plasmids. The promoter fragment (-299 to +35 relative to the CAP site) was fused to the Hind III site of the pSVo plasmid (Gorman et al., 1982) by Hind III linkers; 20 μgs of normal, -175 HPFH or pSVo plasmids were transfected into K562 cells by electroporation; cells were subsequently grown for 42 hours.

The duplicated CCAAT region

The CCAAT motif is an essential element of many eukaryotic promoters, and is characteristically duplicated in the human γ-globin promoter; the transcriptional factor CP1 (Chodosh et al., 1986) has been shown to be able to interact (in vitro) with the proximal and (with three -fold lower affinity) with the distal CCAAT motif of the γ-globin promoter (Superti-Furga et al., 1988). The vertebrate homologue of the sea urchin CCAAT displacement protein (Barberis et al., 1987) also interacts with the CCAAT box region of the γ-globin promoter (Superti-Furga et al., 1988); it is not known whether the interaction of CP1 and CDP in this region is mutually exclusive, as in the sea urchin. Between the two CCAAT boxes, a recognition site (GACAAGG) for NFE-1 also exists (Superti-Furga et al., 1988); in addition, Mantovani et al. (1988), also demonstrated a further erythroid specific protein (band C2 in their paper), which will be referred to as NFE-2. Binding of NFE2 does not require the GACAAGG motif (Mantovani et al., 1988); on the other hand, an oligonucleotide (20-mer) truncated at position -113 is unable to bind NFE-2, locating the 3' border of the NFE2 site between positions -113 and -101. Preliminary data from DMS interference and site directed mutation analysis indicate the importance for binding of the GCCTTG motif that is repeated four times in this region; the two repeats immediately flanking the CCAAT box are the most important for binding.

HPFH due to a G$\longrightarrow$A mutation at position -117, immediately upstream to the CCAAT box, has been described in Greeks and, at high frequency, in Northern Sardinians (Collins et al., 1984; Gelinas et al., 1984; Ottolenghi et al., 1988 a). The combined data (Table I) of Mantovani et al. (1988) and Superti-Furga et al. (1988) indicate that this mutation increases the binding of CP1 (to the distal CCAAT box) and of CDP, while greatly decreasing those of NFE1 and NFE2. These data suggest that either NFE-1 or NFE-2 or both might be repressors of γ-globin gene activity, when bound in the CCAAT box region; the increased binding of CP1, a known transcriptional activator, might additionally contribute to the HPFH phenotype.

Recently, however, HPFH due to a 13 nt. deletion in the $^A\gamma$-globin CCAAT box region has been detected (Gilman et al., 1988 b); binding data obtained with this mutant do not agree with the previous hypotheses. Using a labelled 29 mer corresponding to the -13 nt HPFH sequence, no binding of either CP1 or NFE-2 is demonstrated (data not shown). In agreement with these data, Figure 3A shows that the unlabelled HPFH 29mer is unable to compete against labelled 42mer (corresponding to the undeleted normal sequence) for generating bands C1 (CP1-binding) and C2 (NFE-2 binding); in contrast, an unlabelled oligonucleotide carrying the -117 HPFH mutation competes for CP1, but not NFE-2 binding (see lanes 4 and 6). NFE1 binding cannot be precisely evaluated in this way as an artefactual band (C3 in Mantovani et al., 1988) migrates close to the NFE-1 band; we therefore used unlabelled normal 42-mer and HPFH 29-mer in competition experiments against labelled -175 HPFH oligonucleotide. Figure 3 B shows very similar competition data for normal and -13 nt HPFH oligonucleotides, indicating that the HPFH deletion only slightly, if at all, (no more than two fold in a series of experiments) decreases NFE-1 binding.

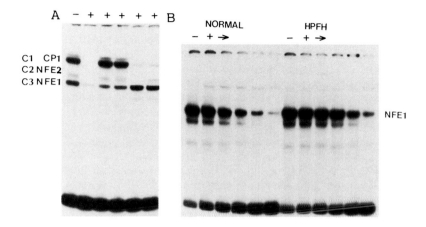

Figure 3. Competition of normal and 13 nt deletion HPFH oligonucleotides against labelled normal CCAAT region 42-mer (A) and -175 HPFH oligomer (containing a high affinity NFE1 binding site) (B).

A. labelled CCAAT box region 42-mer (nucleotides -132 to -91 in Figure 1) was incubated with uninduced K562 cell extract in the absence of competitor (lane 1) or in the presence of a 100-fold excess of unlabelled 42-mer (lane 2), 20-mer (nucleotides -132 to -113) (lane 3), deletion HPFH 29-mer (deleted oligomer spanning the same region as the 42-mer) (lane 4), normal CCAAT box region oligomer (Mantovani et al., 1988; nucleotides -127 to -101) (lane 5) and -117 HPFH CCAAT box region (as above, -117 G→A) (lane 6). Bands are labelled C1, C2, C3 according to Mantovani et al., 1988, and the corresponding factors are indicated; with the 42-mer, band C3 is contaminated with a spuriuos band (Mantovani et al., 1988).

B: labelled -175 HPFH oligomer (Mantovani et al., 1988) was incubated in the absence of competitor (lanes 1 and 7) or with excess unlabelled normal (lanes 2-5) and 13 nt. deletion HPFH (lanes 8-12) 42-mer and 29-mer, respectively, at molar excesses of 50, 100,200, 400, 800-fold.

--: no competitor; +:competitor, as indicated.

DISCUSSION

A note of caution is necessary before attempting any interpretation of the present data; the HPFH phenotype is expressed in adult erythroid cells and it is therefore possible that some of the nuclear proteins from K562 fetal-embryonic cells are not present in adult cells, while other adult stage-specific proteins might be missing in K562 cells. However, the functional data reported in figure 2 reproduce, in part, the difference in γ-globin promoter activity expected for the -

-175 HPFH mutation relative to normal suggesting that the binding alterations detected in this work may be relevant to the HPFH phenotype. The increased activity of the -175 HPFH promoter-correlates with the slightly increased NFE-1 binding activity detected (Figure 2), while decreased OTF-1 binding (also an effect of the -175 mutation) appears to have, per se, little or no functional consequence. Thus, we believe that NFE-1 may be an activator of γ-globin transcription. This factor binds also to the human β-globin enhancer (Mantovani et al., 1988; Wall et al., 1988) and is present in adult mouse erythroid cells (Superti Furga et al., 1988; Wall et al., 1988); therefore we speculate that increased binding of NFE-1 to the -175 HPFH promoter in adult cells might be responsible for the HPFH phenotype. Functional data for the -13 nt. HPFH promoter are not yet available, although slightly increased activity of the -117 HPFH promoter in K562 cells has been reported (Collins et al., this meeting); thus, we can speculate on the function of different nuclear proteins only on the basis of their binding to the different HPFH promoters. Table I reports binding data for CP1, CDP, NFE-1 and NFE-2 in the -117 (Mantovani et al., 1988; Superti-Furga et al., 1988) and -13nt. deletion HPFH'S; CDP binding data for the latter HPFH (not shown in this paper) have been obtained in collaboration with Giulio Superti-Furga (Mantovani et al., 1989).

TABLE I. Effects of the -117 G → A mutation (Greek HPFH) and of the 13 nt. deletion (Black HPFH) on binding of various nuclear factors to the CCAAT box region of the γ-globin promoter.

	CP1	CDP	NFE1	NFE2
-117 HPFH	›(3-fold)	›(2 fold)	«(8-10 fold)	«(10 fold)
13 nt. deletion HPFH	0	0	‹	0

Binding data for CP1 refer to the distal CCAAT box. Data are from Mantovani et al., 1988, 1989 and Superti-Furga et al., 1988.

If we assume that the in vivo HPFH phenotype reflects the binding alterations observed, NFE-1 or NFE-2 (or both), but not CDP, might act as repressors of γ-globin promoter activity in the -117 HPFH; conversely, in the -13 nt. deletion HPFH, NFE-2 and/or CDP, but not NFE-1, might have repressor activity. Taken together the data suggest NFE-2 as the most likely candidate for a repressor protein; while we cannot rule out the possibility that loss of NFE1 and CDP binding has some role in generating the HPFH phenotype of the -117 and -13 nt. deletion mutants, respectively, it is clear that HPFH can occur even in the presence of essentially normal binding of these putative repressors. Mutational analysis of the CCAAT box region is underway to define mutations capable of singly affecting the binding of each of the four proteins discussed above; together with functional analysis of the mutated promoters, it might allow to define precisely the role of these nuclear factors.

ACKNOWLEDGEMENTS

This work was partially supported by: CNR grant 87.00866.51 Progetto Finalizzato Ingegneria Genetica e Basi Molecolari delle malattie Ereditarie to S.O. and by National Institutes of Health Research Grants No. DK35443 (to J.G.G.).

REFERENCES

Barberis A, Superti-Furga G ,Busslinger M (1987). Mutually exclusive interaction of the CCAAT-binding factor and of a displacement protein with overlapping sequences of a histone gene promoter. Cell 50:347-359

Chodosh LA, Baldwin AS, Carthew RW, Sharp PA (1988). Human CCAAT binding proteins have heterologous subunits. Cell 53:11-24

Collins FS, Metherall JE, Yamakawa M, Pan J, Weissman SM, Forget BG (1985). A point mutation in the $^A\gamma$ globin gene promoter in Greek Hereditary persistence of fetal hemoglobin. Nature 313:325-326

Fletcher C, Heintz N, Roeder RG (1987). Purification and characterization of OTF$_1$, a transcription factor regulating cell cycle expression of a human histone H2B gene. Cell 51:773-778

Gelinas R, Endlich B, Pfeiffer C, Yagi H, Stamatoyannopoulos G (1985). G to A substitution in the distal CCAAT box of the $^A\gamma$ globin gene in Greek hereditary persistence of fetal hemoglobin. Nature 313:323-324

Gilman JG, Mishima N, Wen XJ, Kutlar F, Huisman THJ (1988) Upstream promoter mutation associated with a modest elevation of fetal hemoglobin expression in human adults. Blood 72:78-81

Gilman JG, Mishima N, Wen XJ, Stoming TA, Lobel J, Huisman THJ (1988) Distal CCAAT box deletion in the $^A\gamma$ globin gene of two blacks with elevated fetal $^A\gamma$ globin. Nucleic Acids Res 16:10635-10642

Gorman CM, Moffat LF, Howard BH (1982). Recombinant genomes which express chloramphenicol acetyltransferase in mammalian cells. Mol Cell Biol 2:1044-1051

Mantovani R, Malgaretti N, Nicolis S, Ronchi A, Giglioni B, Ottolenghi S (1988). The effects of HPFH mutations in the human γ-globin promoter on binding of ubiquitous and erythroid specific nuclear factors. Nucleic Acids Res 16:7783-7797

Mantovani R, Superti-Furga G, Gilman J, Ottolenghi S (1989) A distal CCAAT box deletion in the $^A\gamma$ globin gene in a type of HPFH causes loss of binding of NFE2, CP1 and CDP, but not NFE-1, nuclear factors. Submitted for publication.

Ottolenghi S, Camaschella C, Comi P, Giglioni B, Longinotti M, Oggiano L, Dore F, Sciarratta G, Ivaldi G, Saglio G, Serra A, Loi A, Pirastu M (1988). A frequent $^A\gamma$-Hereditary persistence of fetal hemoglobin in northern Sardinia: its molecular basis and haematological phenotype in heterozygotes and compound heterozygotes with β-thalassemia. Human Genetics 79:13-17

Ottolenghi S, Nicolis S, Taramelli R, Malgaretti N, Mantovani R, Comi P, Giglioni B, Longinotti M, Dore F, Oggiano L, Pistidda P, Serra A, Camaschella C, Saglio G (1988). Sardinian $^G\gamma$ HPFH: a T$\rightarrow$C substitution in a conserved "octamer" sequence in the $^G\gamma$-globin promoter. Blood 71:815-817

Stamatoyannopoulos G, Nienhuis AW (1987). Hemoglobin switching. in The Molecular Basis of Blood Diseases. G. Stamatoyannopoulos, A.W. Nienhuis, P. Leder and P.W. Majerus, eds. (Philadelphia: W.B. Saunders Company), pp. 66-105

Stoming TA, Stoming GS, Lanclos KD, Fei YJ, Altay C, Kutlar F, Huisman THJ (1988) An $^A\gamma$ type of nondeletional hereditary persistence of fetal hemoglobin with a T$\rightarrow$C mutation at position -175 to the Cap site of the $^A\gamma$globin gene. Blood, in press

Surrey S, Del Grosso K, Malladi P, Schwartz E (1988). A single base change at position -175 in the 5' flanking region of the $^G\gamma$ globin gene from a black with $^G\gamma\beta^+$ HPFH. Blood 71:807-810

Superti-Furga G, Barberis A, Schaffner G, Busslinger M (1988). The -117 mutation in greek HPFH affects the binding of three nuclear factors to the CCAAT region of the γ-globin gene. EMBO J 7:3099-3107

Wall L, deBoer E, Grosveld F (1988). The human β-globin gene 3' enhancer contains multiple binding sites for an erythroid-specific protein. Genes Dev 2:1089-1100

Waber PG, Bender MA, Gelinas RE, Kattamis C, Karaklis A, Sofroniadou K, Stamatoyannopoulos G, Collins FS, Forget BG, Kazazian HHJr (1986) Concordance of a point mutation 5' to the $^A\gamma$globin gene with $^A\gamma\beta^+$ hereditary persistence of fetal hemoglobin in Greeks. Blood 67:551-554

Yang KG, Stoming TA, Fei YI, Liang S, Wong SC, Masala B, Huang KB, Wei ZP, Huisman THJ (1988). Identification of base substitutions in the promoter regions of the $^A\gamma$- and $^G\gamma$-globin genes in $^A\gamma$- (or $^G\gamma$-) β^+-HPFH heterozygotes using the DNA-Amplification-Synthetic Oligonucleotide procedure. Blood 71:1414-1417

Hemoglobin Switching, Part A: Transcriptional Regulation, pages 237–246

A HUMAN GAMMA GLOBIN GENE VARIANT BINDS SP1

Kathryn Sykes and Russel Kaufman

Departments of Biochemistry,(K.S. and R.K.) and
Medicine, (R.K.), Duke University, Durham,
North Carolina 27710

INTRODUCTION

In vitro DNA-protein binding assays have shown that a
300 bp DNA segment located 5' to the human ^{A}gamma globin
gene specifically interacts with factors present in nuclear
extracts of fetal stage erythroleukemia cells (Lingrel
et.al.,1987; Mantevani et.al.,1987). Several classes of
naturally occurring gamma globin gene mutations localized to
this region are associated with elevated fetal and reduced
adult globin gene expression in adults. For example, a
single base substitution 202 bp upstream from the cap site
of the ^{G}gamma globin gene is correlated with an
approximately 40-fold increase in ^{G}gamma globin gene
expression during adult life. In all known patients with the
-202 mutation, only the expression of the ^{G}gamma gene and
not the normal, cis linked ^{A}gamma globin gene is affected.
It is the only non-polymorphic change from the normal
sequence over a 1500 bp range (Collins et.al.,1984). The
mechanism by which this point mutation may lead to the adult
stage overproduction of ^{G}gamma in adults is unknown;
however, several hypotheses have been presented to explain
this clinical condition known as hereditary persistance of
fetal hemoglobin (HPFH). We have directed our studies of
globin gene control to the DNA region 200 bp upsteam from
the cap site.

We now describe factors that interact with this region

in a general or differentiation stage specific manner. Furthermore, we demonstrate that the mutation at position -202 present in the HPFH globin variant produces more avid binding by the general transcription factor, Sp1. We propose that the molecular basis of this HPFH gamma globin variant may relate to the emergence of stage non-specific Sp1 binding.

RESULTS

Identification of Nuclear Factors that Interact with the G-Gamma Globin Gene Upstream Sequences.

To evaluate the presence of specific 5' protein binding sites in the ^{G}gamma globin gene, electrophoretic mobility shift assays were performed. Analysis of naturally occurring gamma globin gene mutations suggested a regulatory function for a DNA region 200 bp upstream from the transcription start site (Collins et.al.,1984). Based on these data we chose to focus our investigation on the sequence immediately surrounding the -200 region. A 27 bp oligonucleotide, gamma 27^{WT}, was prepared that represents 23 bp of gamma globin sequence extending from nucleotide -213 to -191. Four additional bases were added at one end to create restriction sites for experimental manipulations (Table 1).

Nuclear extracts from MEL, K562, and HeLa cells were incubated with the end-labeled synthetic gamma-27^{WT} in order to detect specific nuclear protein binding sites within the -200 upstream region. In all reactions, 1.5 ug (at least 3,000-fold excess) of the heteropolymer poly(dI-dC):poly(dI-dC) was sufficient to inhibit most non-specific DNA-binding. To verify the specificity of the interactions a 50-fold excess of unlabeled synthetic gamma-27^{WT} was used as a competitor for nuclear factors binding to radiolabeled gamma-27 ^{WT}. The products of the DNA-protein incubations were subsequently fractionated on a non-denaturing polyacrylamide gel and analyzed by autoradiography (Fig. 1A). Reproducible patterns of complex formation were obtained.

Five bound complexes are observed when the

radiolabeled gamma-27^{WT} binds protein extracts from
uninduced, adult-stage, mouse erythroleukemia (MEL) cells
and uninduced, fetal-stage, human erythroleukemia (K562)
cells, (Fig. 1A, lanes 2 and 4). The fetal-stage K562 and
adult-stage MEL cells were induced to differentiate, which
permits them to express either fetal or adult globin genes,
respectively. Nuclear proteins from the induced cells were
extracted, incubated with labeled gamma-27^{WT}, and analyzed
by mobility shift. Comparison of the gamma-27^{WT} binding
patterns from induced and uninduced cells demonstrates a
reduction in complex B1 intensity in differentiated
relative to undifferentiated cell extracts (Fig. 1A, lanes
1 and 3 to lanes 2 and 4).

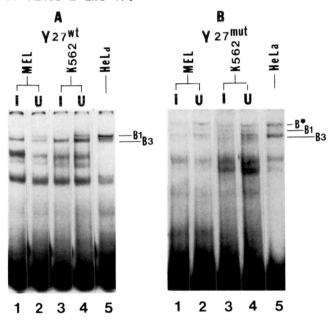

FIG. 1. (A) Specific complexes are formed between the gamma-
27^{WT} probe DNA and nuclear proteins extracted from MEL,
K562, and HeLa cells. (B) The mutated globin sequence,
gamma-27^{MUT}, and nuclear proteins form a unique complex.
Induced cell extracts were produced from cells exposed to 2%
DMSO (MEL cells) or 50 uM hemin (K562 cells) for 56 hours.
Each binding reaction contained approximately .25 ng of
probe, 15 ug of nuclear extract, and 1.5 ug of poly(dI-dC):
poly(dI-dC). I, induced cell extract; U, uninduced cell
extract; B*,B1,B3, protein bound complexes.

The band intensity of complex B1 may appear diminished rather than eliminated in the differentiated cell extracts as a consequence of the incomplete chemical induction process. The inducing agents used for this process produce non-uniform cell differentiation, resulting in a culture population containing a continuum of cells in various differentiated states (Marks et.al.,1978). Hence, most but not all of the cells within the induced culture would have eliminated this interaction.

Complexes B3 and the faster mobility complexes appear in each extract. However, since these fast complexes increase in intensity with prolonged extract storage we believe that these are degradation products of B1 and B3. In addition to the five complexes observed in MEL and K562 cell extracts, the HeLa cell nuclear factors also interact with labeled gamma-27^{WT} to produce another complex. However, the inefficient competition for the factors in this complex by specific unlabeled DNA and its variable appearance suggest that the interaction may be non-specific. Alternatively, its aberrant competition pattern may reflect differences in the binding properties and stoichiometry of the factor(s) in this complex relative to those in other complexes.

The HPFH Gamma Globin Mutation at Position −202 is Associated with an Altered Protein Binding Pattern.

To further characterize the interaction within the 23 bp upstream gamma globin region and to investigate the molecular basis of this naturally occurring HPFH mutant, we synthesized a mutated gamma globin oligonucleotide. Gamma-27^{MUT} differs from gamma 27^{WT} at only one nucleotide position; it contains the same C to G substitution at position −202 which is correlated with elevated adult-stage expression of the ^{G}gamma (fetal) globin gene. We then performed mobility shift assays to determine if the mutation is associated with a change in the nuclear protein binding pattern.

End-labeled gamma-27^{MUT} was incubated with extracts from MEL, K562, and HeLa cells in the presence of a non-specific competitor DNA as described earlier. A comparison of the mobility shift patterns produced by the wild type gamma globin -200 region oligonucleotide to that produced by the mutated oligonucleotide (gamma-27^{WT} and gamma-27^{MUT}) demonstrates that they interact differently with the nuclear proteins (Fig. 1B). Incubation of labeled gamma-27^{MUT} with nuclear proteins reveals the gain of a new low mobility complex, B*, when assayed by mobility shift. The interaction represented by complex B* is present in the nuclear extracts of all cell types tested and the complex is distinct from the other observed complexes.

The Similarities of the Gamma Globin Upstream Region to the SV40 Sp1 Recognition Sites.

The upstream gamma globin sequence centered about -200 relative to the cap site is similar to the sequence recognized by the general transcription factor, Sp1. The C to G base substitution at position -202, correlated with HPFH, increases the similarity of the region to the Sp1 recognition and binding site such that it matches the decanucleotide consensus in 7 out of 10 positions (Table 1). Consequently, it has been suggested that the -202 mutation might produce a novel Sp1 binding site and stimulate transcription (Collins et.al.,1984). To examine this possibility, we characterized the binding competition properties of the gamma globin sequences relative to known Sp1 binding sites. We labeled the SV40 high affinity Sp1 binding site V, SV-V, and incubated it with induced K562 cell extracts. Unlabeled SV40 and gamma globin DNA competitors were added and binding was assayed by mobility shift (Fig.2). The specificity of the SV-V interaction with nuclear extracts is established with an excess of unlabeled SV-V. Relative to SV40 site V, binding site I (SV-I) is known to be a low affinity recogniton site (Gidoni et.al.,1985). A synthetic oligonucleotide representing a mutated site V, SV-X, contains a transversion mutation in the consensus position 4 that dramatically reduces Sp1 binding activity (Jones et.al.,1986). Both SV-I and SV-X compete very weakly for factors binding to the labeled SV-V. These patterns are consistent with the anticipated Sp1 binding specificity (Fig. 2, lanes 2-5). When the gamma globin upstream sequences are added to the incubations as specific unlabeled competitors, we find that the wild type gamma globin sequence (gamma-27^{WT}) serves as a weak competitor while the mutated sequence (gamma-27^{MUT}) is a relatively strong competitor for SV-V binding factors (Fig. 2, lanes 6 and 7).

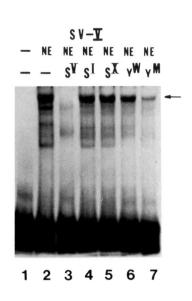

SV-Ⅴ

— NE NE NE NE NE NE

— — s^V s^I s^X γ^W γ^M

1 2 3 4 5 6 7

FIG.2. The mutated gamma globin and the SV40 sequences compete for the same nuclear factors. The specific competition for factors binding to a high affinity Spl binding site (Site V) is analyzed by mobility shift. SV-V, Spl binding site V used as probe; NE, presence of induced K562 cell nuclear extract; s^V,s^I,s^X,gamma^W,gamma^M, unlabeled specific competitor DNA in 50 fold excess over probe; gamma^M, gamma-27^{MUT}; gamma^W, gamma-27^{WT}; s^V, SV-V; s^I, SV-I; s^X, SV-X; arrow indicates position of complex competed off the probe DNA by both SV40 and globin sequences.

Similar binding affininty results were obtained by labelling the mutated gamma globin DNA and competing with SV40 SP1 binding site oligonucleotides (data not shown). Therefore, we conclude that, in vitro, the mutated gamma globin -200 region is able to compete moderately well with factors that bind an SV40 Spl binding site. In addition, we note that the low mobility complex which is observed bound to SV-V migrates similarly with a complex bound to gamma-27^{MUT} (Fig. 3, lanes 1 and 6). These data are consistent with the hypothesis that the mutated upstream globin sequence binds Spl or an Spl-like factor. In addition to using the induced K562 cell extracts, the experiments were repeated using uninduced K562 and HeLa cell nuclear extracts and similar competition results were obtained.

Purified Sp1 Binds the Gamma Globin Upstream Region In Vitro.

To demonstrate more conclusively that the globin sequences interact with the general transcription factor, Sp1, we performed similar mobility shift binding studies with a preparation of 60-90% pure Sp1 (a gift of S. Jackson and R. Tjian). Radiolabeled gamma globin upstream sequences (gamma-27^{WT}, gamma-27^{MUT}) were each incubated with the purified protein and binding was analyzed in non-denaturing gels. Figure 3 shows that purified Sp1 forms a specific complex with both the wild type and mutated DNA fragments. The mutated sequence exhibits a greatly enhanced binding activity relative to the wild type sequence (Fig. 3, lanes 5 and 8); however, its activity is somewhat reduced relative to one of the highest Sp1 affinity sites (data not shown). We find that the Sp1 complex is competed off the labeled gamma globin DNA with an excess of SV40 Sp1 binding

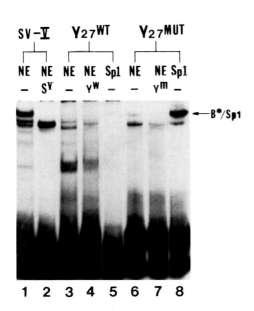

FIG.3. Purified Sp1 protein binds the gamma globin upstream sequences. The interactions of the SV40 and gamma globin sequences with HeLa cell nuclear extracts and purified Sp1 are compared by mobility shift. SV-V, gamma-27^{WT},gamma-27^{MUT}, synthetic DNAs used as probes; NE, nuclear extract; S^{V},gamma^{W},gamma^{M}, unlabeled specific competitor DNA in 50 fold excess over probe; S^{V}, SV-V; gamma$^{W}_{M}$, gamma-27^{WT}; gamma^{M}, gamma-27^{MUT}; B*, novel low mobility complex.

site V (SV-V) but not the mutated binding site (SV-X) (data not shown). The gamma 27^{MUT}-Spl complex migrates similarly with the novel low mobility complex, B*, observed when either labeled gamma-27^{MUT} or SV-V are incubated with crude nuclear extracts from the HeLa cells (Fig. 3, lanes 1, 6, and 8). Taken together, these data demonstrate that under our DNA excess, in vitro binding conditions, the general transcription factor Spl binds sequences 200 bp upstream from the ^{G}gamma globin gene cap site. The relative binding activities of the SV40 and globin sequences are summarized in Table 1.

TABLE 1. COMPARISON OF SV40 AND G GAMMA GLOBIN SEQUENCES TO THE CONSENSUS SPl RECOGNITION AND BINDING SITE

DNA	Recognition Site[a,b,c]	Relative Spl Binding Activity
	5' 3'	
gamma 27^{WT}	-213 -191 aat tCT CTT GGG GGC CCC TTC CCC ACA	+/-
gamma 27^{MUT}	* aat tcT CTT <u>GGG GGC GCC</u> T<u>T</u>C CCC ACA	++

SP1 Consensus	G G GC
	          G GGC GG A AT
	          T	

SV-V	-81 -92 aat tcG ATG GGC GGA GTa	++++
SV-I	-38 -51 aat tCA TGG GGC GGA GAA	+
SV-X	-81 -92 tc gaG ATG GCC GGA GTt c	+/-

[a] Top strand of double stranded oligonucleotide is given.
[b] Globin and SV40 nucleotides are capitalized; restriction site linker sequences are lower case. Numbers above gamma globin sequences designate nucleotide position relative to the start site; numbers above SV40 sequences designate nucleotide positions according to Tooze, 1981.
[c] Solid lines define postions of identity with the optimal consensus Spl site; the dot identifies a position of identity with a weak Spl binding site, HSV-TK(I) (Jones,1985). The asterisk identifies the -202 mutation site.

DISCUSSION

In this study we use mobility shift analysis to demonstrate nuclear protein binding sites arranged within a 23 nucleotide stretch (-213 to -191 bp) upstream of the human ^{G}gamma globin transcription unit. Nuclear extracts from undifferentiated cells form a specific complex (B1) with the upstream region. Differentiated erythroleukemia cell extracts form a similar complex (B1), but it appears with reduced intensity in mobility shift assay. In contrast, complex B3 is detected in each of the extracts tested. Still another complex (B*) is associated only with the gamma globin upstream sequence when it contains a single base substitution at nucleotide position -202. Sequence comparisons and in vitro binding experiments using purified protein have permitted us to conclude that the general transcriptional activator Sp1 binds the mutated DNA region. The results suggest that complex B* contains Sp1.

Several hypotheses have been presented to explain the adult-stage overproduction of ^{G}gamma globin, associated with the -202 mutation. First, the mutation may lead to constitutive expression by disrupting the function of a negative regulatory element. Second, the -202 mutation may augment transcription by increasing the efficiency of a positive regulatory element normally spanning the region. Last, the mutated sequences may stimulate transcription by creating a novel positive regulatory element, absent from the wild type sequence (Collins et.al.,1984). The results we present support the last of these possibilities.

ACKNOWLEDGMENTS

This work was supported by Public Health Service grants DK38699 and HL-23891 from the National Institutes of Health. R.E.K. is a Scholar of the Leukemia Society of America.

LITERATURE CITED

Collins, F., C. Stoeckert, G. Serjeant, B. Forget, and S. Weissman (1984). ^{G}gamma beta$^+$ hereditary persistence of fetal hemoglobin: cosmid cloning and identification of a specific mutation 5' to the ^{G}gamma gene. Proc. Natl. Acad. Sci. USA 81:4894-4898.

Collins, F. and B. Forget (1985). A point mutation in the ^{A}gamma globin gene promoter in the Greek type of hereditary persistance of fetal hemoglobin. Nature (London) 313:325-326.

Dyan, W. and R. Tjian (1983). The promoter-specific transcription factor Sp1 binds to upstream sequences in the SV40 early promoter. Cell 35:79-87.

Gidoni, D., J. Kadonaga, H. Barrera-Soldana, K. Takahashi, P. Chambon, R. Tjian (1985). Bidirectional SV40 transcription mediated by tandem Sp1 binding interactions. Science 230:511-517.

Jones, K., K. Yamamoto, and R. Tjian (1985). Two Distinct transcription factors bind to the HSV thymidine kinase promoter in vitro. Cell 42:559-572.

Jones, K., J. Kadonaga, P. Luciu, and R. Tjian (1986). Activation of the AIDS retrovirus promoter by the cellular transcription factor, Sp1. Science. 232:755-759.

Lingrel, J., J. Weimer, and A. Menon (1987). Binding of nuclear factors to an upstream region of the human ^{A}globin gene, pp. 201-210. In G. Stamatoyannopoulos and A. Nienhuis (eds.), Developmental Control of Globin Gene Expression. Alan R. Liss, New York.

Mantevani, R. N. Malgaretti, B. Giglioni, P. Comi, N. Cappellini, S. Nicolio, S. Ottolenghi. (1987). A protein factor binding to an octamer motif with gamma globin promoter disappears upon induction of differentiation and hemoglobin synthesis in K562 cells. Nucleic Acids Res. 15:9349-9364.

Marks, P. and R. Rifkind. (1978). Erythroleukemic Differentiation. Ann. Rev. Biochem. 47:419-448.

Tooze, J.(1981). Molecular Biology of Tumor Viruses, 2nd ed., part 2, revised. DNA Tumor Viruses. Cold Spring Harbor Laboratory, Cold Spring Harbor, New York.

Hemoglobin Switching, Part A: Transcriptional Regulation, pages 247–260
© 1989 Alan R. Liss, Inc.

NUCLEAR PROTEINS OF A HUMAN ERYTHROLEUKEMIC CELL LINE THAT
BIND TO THE PROMOTER REGION OF NORMAL AND NONDELETION HPFH
γ-GLOBIN GENES

James E. Metherall, Frances P. Gillespie and
Bernard G. Forget

Departments of Human Genetics and Internal
Medicine, Yale University School of Medicine,
New Haven, Ct. 06510

INTRODUCTION

Naturally occuring mutations exist that interfere with
the process of fetal to adult hemoglobin switching (for
reviews see Weatherall and Clegg, 1981; Collins and
Weissman, 1984; Stamatoyannopoulos and Nienhius, 1987; Bunn
and Forget 1986). These mutations result in a condition
known as hereditary persistence of fetal hemoglobin (HPFH).
Adults affected with HPFH demonstrate elevated levels of
fetal γ chain production and decreased levels of adult β
chain production. Nondeletion forms of HPFH (ndHPFH) have
been identified and are typically characterized by selective
overexpression of one of the two γ globin genes (^{G}γ or ^{A}γ).
The molecular analysis of a number of forms of ndHPFH has
led to the identification of single base mutations within
the transcriptional promoters of the overexpressed γ globin
genes. It has been proposed that these mutations affect the
binding of transcriptional regulatory factors and allow
continued expression of the affected γ gene in adult life.

A point mutation at position -117 of the ^{A}γ globin gene
is associated with one form of Greek ^{A}γ ndHPFH (Collins et
al., 1985; Gelinas et al., 1985; Metherall et al., 1988).
This mutation is located 2 nucleotides 5' of the more distal
of the duplicated CCAAT boxes of the γ genes. Aberrant
expression of the ndHPFH ^{A}γ gene in the human
erythroleukemic cell line KMOE has been attributed to this
single base change (Stoeckert et al., 1987). KMOE cells
normally produce low but detectable levels of both γ and β
globin mRNA. However, when KMOE cells are treated with

cytosine arabinoside (araC) the level of β mRNA increases subtantially while $^A{}_{γ}$ mRNA levels are unaffected. The normal and position -117 $^A{}_{γ}$ genes were introduced into these cells using a retroviral expression system. Cells harboring the position -117 mutant gene demonstrated markedly increased levels of γ mRNA following araC treatment, whereas cells harboring the normal $^A{}_{γ}$ gene showed no increase in the amount of γ mRNA (Stoeckert et al., 1987). Although the mechanism by which araC affects globin gene expression is not well understood, the finding that the normal and position -117 $^A{}_{γ}$ genes are differentially expressed in araC treated KMOE cells suggests that these cells contain a transcriptional regulatory factor that can distinguish between the normal and position -117 promoters.

Using a nitrocellulose blot protein-DNA binding assay, we have identified a number of γ promoter binding activities in KMOE cells. In extracts prepared from araC treated cells specific high affinity binding activities were observed at 76 kDa, 65 kDa and 36 kDa. While the 36 kDa activity was also observed in extracts prepared from untreated cells, the 76 kDa and 65 kDa activities were not observed. All three proteins bound equally well to the normal and position -117 promoters. At higher DNA concentrations, a large number of lower affinity interactions were observed. These interactions included a 72 kDa protein that bound to the normal promoter but not to the position -117 ndHPFH promoter. The correlation between γ gene expression and the binding activity of the 72 kDa protein in these cells suggests that this protein may be a transcriptional repressor that normally binds to the γ gene promoter during adult life to prevent transcription. A mutation in the binding site of this presumed repressor may explain the specific overexpression of the $^A{}_{γ}$ gene in adults affected with Greek $^A{}_{γ}$ ndHPFH.

RESULTS

Proteins Induced by araC Bind to the γ Gene Promoter

Nuclear proteins were prepared from KMOE cells by the method of Dignam et al. (1987), resolved by polyacrylamide gel electrophoresis, transferred to nitrocellulose and incubated with radiolabeled DNA probes according to the method of Miskimins et al. (1985). Probes were prepared

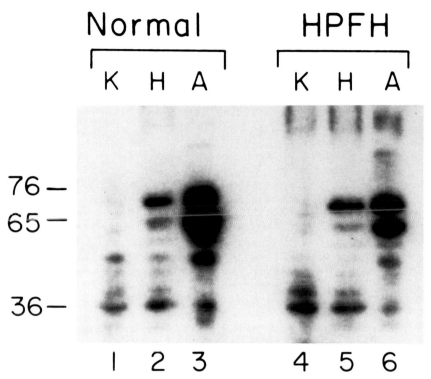

Fig. 1. **DNA binding proteins in KMOE cells.** KMOE cells
were cultured in suspension for 6 days in RPMI medium
containing 5% fetal calf serum (FCS) followed by two days of
culture in medium containing 15% FCS. The cultures were
diluted to 5×10^4 cells/ml and allowed to grow for an
additional 2 days. Cells were then treated for 2 days in
medium containing either no additions (K), 40 μM hemin (H)
or 10 μM araC (A). Total nuclear protein was prepared,
resolved by SDS-polyacrylamide gel electrophoresis and
subjected to nitrocellulose blot protein-DNA binding
analysis (Miskimins et al., 1985) using probes from the
normal (lanes 1-3) or position -117 HPFH (lanes 4-6) γ gene
promoters. Blots were exposed to X-ray film with an
intensifying screen for 3-10 hr at -70°C. The calculated
molecular weights of the major binding activites are
indicated.

from either the normal ɣ gene promoter or from the ɣ gene promoter that contains the position -117 ndHPFH mutation. Probes consisted of end-labeled 1344 base pair fragments extending from a Hind III site at position -1247 to a Bgl I site at position +97 relative to the site of transcription initiation. DNA sequence analysis has demonstrated that the only difference between these two promoter fragments is the position -117 mutation (Stoeckert et al., 1987).

A number of proteins bound to the ɣ gene promoter fragments (Figure 1). In untreated KMOE cells a 36 kDa protein bound to both the normal and position -117 promoter fragments. This binding activity was similar in cells treated with hemin (lanes 2 and 5) and araC (lanes 3 and 6). In contrast, 76 kDa and 65 kDa activities that were observed in araC treated cells (lanes 3 and 6) were not observed in untreated cells (lanes 1 and 4). The 76 kDa and 65 kDa activites were also visible in hemin treated cells (lanes 2 and 5), although at a lower level than in araC treated cells. These proteins bound to both the normal and position -117 (HPFH) gene promoter fragments (lanes 1-3 and 4-6).

Specificity of Binding to the ɣ Gene Promoter

Competition analyses were performed in order to determine the specificity with which the 76 kDa, 65 kDa and 36 kDa proteins bound to the ɣ gene promoter (Figure 2). Strips of nitrocellulose containing electrophoretically separated nuclear proteins from araC treated KMOE cells were probed with a normal ɣ gene promoter fragment that had been mixed with increasing amounts of either unlabeled ɣ gene promoter fragment as a specific competitor (closed symbols) or unlabeled salmon sperm DNA as a non-specific competitor (open symbols). The intensities of the signals produced were quantitated using a scanning densitometer and plotted in a double log plot against the weight ratio of competitor DNA to probe DNA. The ɣ promoter fragment was a much better competitor for binding of all three proteins than was the salmon sperm DNA, indicating that the binding of the 76 kDa, 65 kDa and 36 kDa proteins is specific for the ɣ gene promoter.

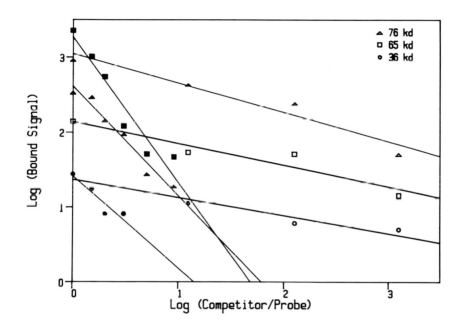

<u>Fig. 2.</u> **Specificity of DNA binding proteins in araC treated**
KMOE cells. Nuclear proteins were prepared from araC
treated KMOE cells and subjected to nitrocellulose blot
protein-DNA binding assays as described in the legend to
Figure 1. Probes consisted of a mixture of labeled normal γ
gene promoter fragment and unlabeled competitor DNA.
Specific competitor DNA consisted of unlabeled γ promoter
fragment (closed symbols) and non-specific competitor
consisted of salmon sperm DNA (open symbols).
Autoradiographic signals for the 76 kDa (triangles), the 65
kDa (squares) and 36 kDa (circles) proteins were quantitated
by densitometry and are presented as the log of the signal
verses the log of the weight ratio of competitor DNA to
probe DNA.

Competition Between the Normal and Position -117 Promoter Fragments

In light of the finding that the normal and position -117 ɣ genes were differentially expressed in araC treated KMOE cells, it was of interest to determine whether the araC-induced proteins bound differentially to the two promoter fragments. Although it is apparent that the proteins bind to both the normal and position -117 promoter fragments (Figure 1), subtle differences in the affinities of these interactions could not be ruled out.

To address whether subtle differences in binding affinity exist between the normal and position -117 promoter fragments, competition studies were performed (Figure 3). Labeled promoter fragment was combined with unlabeled promoter fragment to produce a series of probe mixtures with labeled/unlabeled mole fractions varying from 1.0 to 0.2. These mixtures were then used to probe strips of nitrocellulose filters to which electrophoretically separated nuclear proteins from araC treated KMOE cells had been transferred. The first set of five lanes demonstrates the ability of unlabeled normal promoter fragment (xN) to compete for the binding of labeled normal promoter fragment (N). The second set of five lanes demonstrates the ability of unlabeled HPFH promoter fragment (xH) to compete for the binding of labeled normal promoter fragment (N). And the final set of five lanes represents the reciprocal experiment in which we assesed the ability of unlabeled normal promoter fragment (xN) to compete for the binding of the labeled HPFH promoter (H) fragment.

The results of this competition experiment demonstrate that there is no significant difference in the binding of the araC-inducible proteins (76 kDa and 65 kDa) to the normal and position -117 promoter fragments. This has been confirmed by quantitative tracings of the signals shown in Figure 3 (data not shown). The 36 kDa protein also binds equally well to both promoters as does a fourth protein of 12 kDa. The 12 kDa protein appears to be non-specific, since it is effectively competed by salmon sperm DNA, poly d(I-C) and bacterial plasmid DNA (data not shown).

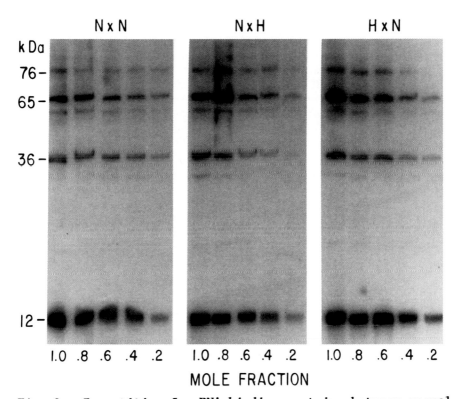

Fig. 3. **Competition for DNA-binding proteins between normal and position -117 promoter fragments.** Nitrocellulose blot protein-DNA binding assays were performed on araC treated KMOE nuclear protein as described in the legend to Figure 2. Probes consisted of a mixture of labeled and unlabeled γ gene promoter fragments. In the first panel, labeled normal promoter fragment (N) was competed with increasing amounts of unlabeled normal promoter fragment (xN), with the labeled probe representing the indicated fraction of the mixture. The second panel shows a similar analysis in which the competitor DNA was the position -117 mutant promoter fragment (xH). In the final panel labeled probe was the position -117 promoter fragment (H), and the competitor DNA was the normal promoter fragment (xN). The calculated molecular weights of the major binding activities are indicated.

**Effect of DNA Concentration on Spectrum of Binding proteins
in araC Treated KMOE cells.**

All the nitrocellulose blot protein-DNA binding assays
described above utilized probe concentrations in the range
of 1×10^{-12}M to 1×10^{-11}M. Interactions that occur at such
low DNA concentrations must be of high affinity. In order
to investigate lower affinity interactions, nuclear proteins
from araC treated KMOE cells were incubated with with
increasing concentrations of DNA (Figure 4). Lanes 1-7 and
lanes 8-14 represent two sets of experiments where the
concentration of normal γ promoter fragment was varied in
half log increments from 1×10^{-15}M to $1\times10^{-8.5}$M. Lanes 15-21
show a similar set of experiments using labeled position
-117 promoter fragment over a concentration range of
$1\times10^{-11.5}$M to $1\times10^{-8.5}$M.

As the DNA concentration is increased, the number of
different proteins that bind increases due to lower affinity
interactions that occur at higher DNA concentrations. The
highest affinity interactions can be seen at concentrations
greater than $1\times10^{-12.5}$M and include the 12 kDa, the 76 kDa
and the 65 kDa proteins (lanes 4-6). The 36 kDa protein is
observed at concentrations greater than 1×10^{-11}M (lanes
9-14) and an increasing number of proteins are observed at
concentrations greater than 1×10^{-10}M (lanes 12-14).

Interestingly, a protein of 72 kDa appears to bind to
the normal γ promoter fragment with higher affinity than it
binds to the position -117 γ promoter; at concentrations
greater than $1\times10^{-10.5}$M a protein of 72 kDa can be seen in
lanes probed with the normal promoter fragment (lanes
10-13). Although all of the other bands observed in this
concentration range are present in the lanes probed with the
position -117 promoter fragment (lanes 17-20), the 72 kDa
protein band is absent. Competition experiments using DNA
concentrations in this range have confirmed that the 72 kDa
protein has little or no affinity for the position -117
promoter fragment (data not shown).

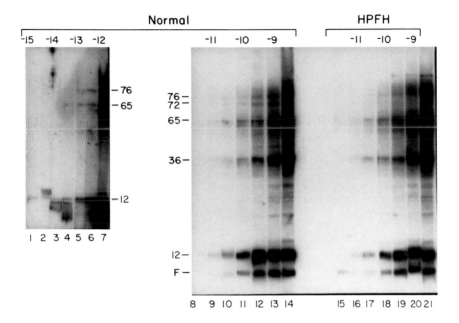

Fig. 4. Effect of DNA concentration on spectrum of
DNA-binding proteins in araC treated KMOE cells.
Nitrocellulose blot protein-DNA binding assays were
performed on araC treated KMOE nuclear protein as described
in the legend to Figure 2.   Probes were prepared from
promoters of either the normal or position -117 (HPFH) γ
genes.   The probe concentrations were varied in half log
increments and molar concentrations are indicated at the top
of every second lane.   The calculated sizes of the major
binding activities are shown along the side of the figure as
is the position of the dye front (F).

DISCUSSION

We have utilized the aberrant expression of a cloned ndHPFH γ gene to investigate the mechanism by which the fetal γ globin genes are silenced in a human adult erythroid cellular environment. This cellular environment is experimentally defined by araC treated KMOE cells. The differential expression of the normal and ndHPFH $^A\gamma$ genes in these cells indicates that the cells contain a transcriptional regulatory factor that can distinguish between the two promoters. Since the only difference between these two promoter fragments is the position -117 mutation (Stoeckert et al., 1987), any difference in expression and/or DNA-protein interactions between these two fragments must be due to this mutation.

The γ globin gene promoters contain a number of recognized transcriptional control elements. These elements include proximal and distal CCAAT box motifs (Breathnach and Chambon, 1981; Graves et al., 1986) located at positions -89 and -115, respectively. The position -117 mutation is located 2 bp 5' of the distal CCAAT box. A number of CCAAT box-binding proteins have been identified which belong to the family known as nuclear factor 1 (NF-1). NF-1 binds to double-stranded sequences containing TGG and its complement (CCA), with the highest affinity binding site being the palindromic sequence, $\underline{TGGN}_7\underline{CCA}$ (Gil et al., 1988a). In addition to the $\underline{CCAAT}$ boxes, there is a potential high affinity binding site for NF-1 located at position -152 ($\underline{TGG}CTAAACT\underline{CCA}$) of the γ gene promoters.

Here we describe a 36 kDa binding activity that demonstrates high affinity for both the normal and position -117 γ gene promoters and is present both before and after araC treatment. This protein may be related to the NF-1 family of transcription factors. Members of the NF-1 family are susceptible to proteolysis during isolation. Although the cloned genes for NF-1/Red1 and NF-1/X predict molecular weights of 62.7 kDa and 48.8 kDa respectively, the binding activities were purified as proteolytic fragments of 33-35 kDa (Gil et al., 1988b). The coincidental size and ubiquitous nature of the 36 kDa protein suggest that it may represent a proteolytic fragment of a member(s) of the NF-1 family. The involvment of NF-1 in γ gene expression is supported by the observation that the position -117 mutation

results in a 2-fold increase in the binding of members of
this family (CP1 and ɣCAAT) to the distal CCAAT box
(Superti-Furga et al., 1988; Gumucio et al., 1988).
However, the ubiquitous nature of NF-1 and the only modest
change in binding affinity argue against a major role of
NF-1 in developmental regulation of ɣ gene expression.

High affinity ɣ promoter-binding proteins of 76 kDa and
65 kDa were also observed. These proteins did not
discriminate between the normal and position -117 promoters
but both proteins were induced to high levels in araC
treated cells. We have found that the 76 kDa and 65 kDa
proteins also bind to the β globin gene promoter (data not
shown). These findings provide a possible explanation for
the araC-mediated induction of β gene expression and suggest
a potential relationship between the 76 kDa and 65 kDa
proteins and a recently described erythroid-specific
DNA-binding protein (NF-E1; Wall et al., 1988; deBoer et
al., 1988). NF-E1 interacts with the β globin gene promoter
as well as the intragenic enhancer and the 3' enhancer.
NF-E1 is expressed in all stages of development and
recognizes the consensus sequence (A/C Py T/A ATC A/T Py).
A sequence matching this consensus can be found at position
-191 of the ɣ gene promoter and proteins (GF-1 and EFɣa)
that interacts with this sequence have recently been
identified (Martin et al., 1988; Gumucio et al., 1988;
Mantovani et al., 1988). The relationship between GF-1,
EFɣa, NF-E1 and the proteins described here remains to be
determined.

Among many ɣ promoter-binding proteins observed at
higher DNA probe concentrations was a 72 kDa protein that
bound to the normal ɣ gene promoter but not to the position
-117 mutant promoter. The correlation between the binding
of this protein and ɣ gene expression suggests that the 72
kDa protein may be involved in repressing ɣ gene
transcription. The inability of the 72 kDa protein to bind
to the position -117 promoter may be responsible for ɣ gene
overexpression in araC treated KMOE cells.

Expression studies have demonstrated that there is no
significant difference between the level of expression of
the position -117 gene and the normal ^{A}ɣ gene in untreated
KMOE cells (Stoeckert er al., 1987). Consistent with this
observation is our inability to detect the 72 kDa protein in
untreated cells (unpublished observations). However, the
finding that other ɣ promoter-binding proteins are induced

by araC confuses this correlation. It is possible that the
low levels of 76 kDa and 65 kDa activity found in untreated
cells are rate-limiting to transcription and that an effect
of the 72 kDa protein can be observed only in araC treated
cells, where these rate-limiting steps have been eliminated.

In gel mobility shift assays a protein has recently
been identified that interacts with the ɣ gene promoter and
shows a 8-10 fold reduction in affinity for the position
-117 promoter (Superti-Furga et al., 1988; Mantovani et al.
1988). This protein contacts each of three copies of an
evolutionarily conserved sequence, (C)CTTGAC, that overlap
the duplicated CCAAT boxes. It has been proposed that this
sequence is involved in developmental control of globin gene
expression since it is conserved only in fetal and embryonic
globin genes (Collins and Weissman, 1984). The position
-117 mutation changes the most 5' of these sequences from
CCTTGAC to CCTTAAC. The relationship between NF-E and this
developmentally conserved sequence is unclear since NF-E
also interacts with the CCAAT box region of the β globin
promoter.

Although far from complete, our current working model
for the regulation of globin gene expression in KMOE cells
involves the following characteristics. Ubiquitous factors,
such as the 36 kDa protein, may be involved in ɣ gene
expression but probably do not play a major role in
distinguishing the normal and position -117 ɣ-globin genes.
AraC induces factors (76 kDa and 65 kDa) that interact with
both the β and ɣ globin gene promoters. These factors may
function as activators of transcription. While induction of
these factors may be sufficient to enhance β gene
transcription, enhancement of ɣ gene transcription may be
prevented by the existence of a repressor protein (72 kDa).
The position -117 ndHPFH mutation may prevent the binding of
this repressor and allow araC-mediated induction of ɣ gene
transcription in KMOE cells. The relationship between these
proteins and known transcription factors remains to be
determined.

ACKNOWLEDGEMENTS

This work was supported in part by grants from the
National Institutes of Health.

REFERENCES

Breathnach R, Chambon P (1981). Organization and expression of eucaryotic split genes coding for proteins. Annu Rev Biochem 50: 349-383.

Bunn HF, Forget BG (1986). "Hemoglobin: Molecular, Genetic and Clinical Aspects." Philadelphia: Saunders.

Collins FS, Weissman SM (1984). The molecular genetics of human hemoglobin. Prog Nucleic Acid Res Molec Biol 31: 315-458.

Collins FS, Metherall JE, Yamakawa M, Pan J, Weissman SM, Forget BG (1985). A point mutation in the Aγ globin gene promoter in Greek hereditary persistence of fetal hemoglobin. Nature (London) 313: 325-326.

deBoer E, Antoniou M, Mignotte V, Wall L, Grosveld F (1988). The human β-globin promoter; nuclear protein factors and erythroid specific induction of transcription. EMBO J 7: 4203-4212.

Dignam JD, Lebowitz RM, Roeder RG (1983). Accurate transcription initiated by RNA polymerase II in a soluble extract from isolated mammalian nuclei. Nucleic Acids Res 11: 1475-1489.

Gelinas R, Endlich B, Pfeiffer C, Yagi M, Stamatoyannopoulos G (1985). G to A substitution in the distal CCAAT box of the Aγ-globin gene in Greek hereditary persistence of fetal hemoglobin. Nature (London) 313: 323-324.

Gil G, Smith JR, Goldstein JL, Slaughter CA, Orth K, Brown MS, Osborne TF (1988a). Multiple genes encode nuclear factor 1-like proteins that bind to the promoter for 3-hydroxy-3-methylglutaryl-coenzyme A reductase. Proc Natl Acad Sci 85: 8963-8967.

Gil G, Osborne TF, Goldstein JL, Brown MS (1988b). Purification of a protein doublet that binds to six TGG-containing sequences in the promoter for hamster 3-hydroxy-3-methylglutaryl-coenzyme A reductase. J Biol Chem 263: 19009-19019.

Graves BJ, Johnson PF, McKnight SL (1986). Homologous recognition of a promoter domain common to the MSV LTR and the HSV tk gene. Cell 44: 565-576.

Gumucio DL, Rood KL, Gray TA, Riordan MF, Sartor CI, Collins FS (1988). Nuclear proteins that bind the human γ-globin gene promoter: Alterations in binding produced by point mutations associated with hereditary persistence of fetal hemoglobin. Mol Cell Biol 8: 5310-5322.

Mantovani R, Malgaretti N, Nicolis S, Ronchi A, Giglioni B, Ottolenghi S (1988). The effects of HPFH mutations in the human ɤ-globin promoter on binding of ubiquitous and erythroid specific nuclear factors. Nuc Acids Res 16: 7783-7797.

Martin DIK, Tsai S, Orkin SH (1988). An erythroid-specific factor binds to the β-globin promoter and mediates expression of an HPFH phenotype. Blood 72 (Suppl 1): 67a.

Metherall JE, Gillespie FP, Forget BG (1988). Analyses of linked β globin genes suggest that non-deletion forms of hereditary persistence of fetal hemoglobin are bona fide switching mutants. Am J Hum Genet 42: 476-481.

Miskimins WK, Roberts MP, McClelland A, Ruddle FH (1985). Use of a protein-blotting procedure and a specific DNA probe to identify nuclear proteins that recognize the promoter region of the transferrin receptor gene. Proc Natl Acad Sci USA 82: 6741-6744.

Stamatoyannopoulos G, and Nienhius AW (1987). Hemoglobin Switching. In Stamatoyannopoulos G, Nienhius AW, Leder P, and Majerus PW (eds.): "The Molecular Basis of Blood Diseases," Philadelphia: W.B. Saunders, pp 66-105.

Stoeckert CJ, Metherall JE, Yamakawa M, Weissman SM, Forget BG (1987). Expression of the affected Aɤ-globin gene associated with nondeletion hereditary persistence of fetal hemoglobin. Mol Cell Biol 7: 2999-3003.

Superti-Furga G, Barberis A, Schaffner G, Busslinger M (1988). The -117 mutation in Greek HPFH affects the binding of three nuclear factors to the CCAAT region of the β-globin gene. EMBO J 7: 3099-3107.

Wall L, deBoer E, Grosveld F (1988). The human β-globin gene 3' enhancer contains multiple binding sites for an erythroid-specific protein. Genes and Dev 2: 1089-1100.

Weatherall DJ, Clegg JB (1981). "The Thalassemia Syndromes." 3rd ed. Oxford: Blackwell Scientific Publications.

Hemoglobin Switching, Part A: Transcriptional Regulation, pages 261–268

PROMOTER SEQUENCE DIFFERENCES MAY DETERMINE MAXIMAL HEMO-
GLOBIN F LEVELS IN 5-AZACYTIDINE TREATED BABOONS

Donald E. Lavelle and Joseph DeSimone

Departments of Medicine and Genetics, University
of Illinois College of Medicine, and the VA
Westside Medical Center, Chicago, Illinois 60612

INTRODUCTION

    The baboon is an excellent experimental model system
to study the control of globin gene expression. Adult
baboon HbF levels increase following erythropoietic stress
caused by phenylhydrazine induced hemolysis, phlebotomy,
or hypobaric hypoxia (DeSimone et al 1978, 1982). Treatment
with the cytotoxic and or cytostatic agents 5-azacytidine
(DeSimone et al 1982), hydroxyurea (Letvin et al 1984), and
cytosine arabinoside (Papayannopoulou et al 1984) also
causes increased HbF levels.

    In baboons, the level of HbF produced in response to
erythropoietic stress (DeSimone et al 1980) or 5-azacyti-
dine (DeSimone et al 1984) treatment is determined by
genetic factors. Among different baboon species treated
with 5-azacytidine, P. anubis is a high HbF responder
(40–80% HbF), P. cynocephalus is an intermediate to low HbF
responder (10–40% HbF), and P. papio and P. hamadryas are
low HbF responders (<20% HbF) (DeSimone et al 1984). Humans
generally produce low levels of HbF when treated with
5-azacytidine, although the individual response varies.
Because a critical minimal HbF level may be necessary to
relieve the symptoms of sickle cell anemia (Perrine et al
1978), our ability to predict an individual's maximal HbF
level in response to various treatments would be important.

    The baboon contains duplicated γ globin genes, denoted
Iγ and Vγ. The Iγ gene codes for isoleucine at position
75 while the Vγ codes for valine at this position. On the

basis of a change in postnatal ratio of Iγ to Vγ, the Iγ
gene appears to be analogous to the human Gγ gene, while
the Vγ gene is analogous to the human Aγ (DeSimone et al
1984). We report here that a G to A base change in P.
anubis at position -82 flanking the proximal CCAAT sequence
of the Iγ and Vγ promoters is correlated with a high HbF
response.

MATERIALS AND METHODS

    Phlebotomized baboons were treated with 5-azacytidine
as previously described (DeSimone et al 1982). HbF levels
were determined by alkali denaturation (Singer et al 1951).
Maximal HbF levels generally occurred 4-5 days following the
last injection of drug. DNA obtained from baboon bone marrow
cells was digested with Bgl II and size fractionated on
sucrose gradients. Phage libraries constructed in the EMBL3
vector (Frischauf et al 1983) were prepared from fractions
containing 10-16 kb DNA. Recombinant phage with the 13 kb
Bgl II fragment containing both fetal globin genes were
identified by hybridization to a human 2.1 kb Bgl II-Bam H1
5'Gγ probe. The baboon Iγ and Vγ genes were each subcloned
in the PTZ18U vector (USBC, Cleveland, Ohio) as 2.1 kb and
5.0 kb fragments, respectively. For sequence analysis
subclones were prepared in M13 vectors (Messing 1983) and
sequencing reactions performed by the dideoxy chain
termination method using DNA polymerase I Klenow fragment
(Sanger et al 1977) and by the chemical method (Maxam and
Gilbert 1980).

RESULTS

    The Iγ and Vγ genes of 6 different baboons were cloned
in EMBL3 phage as 13 kb Bgl II fragments. Three of the
baboons were P. anubis, 2 were P. cynocephalus, and one was
P. papio. The nucleotide sequence of the Iγ and Vγ gene
promoter regions from -300 to +1 was determined in clones
obtained from the six animals. The promoter sequences of the
Iγ genes from individual P. anubis, P. cynocephalus and P.
papio animals are compared to a human Gγ (Shen et al 1981)
sequence in Figure 1. Many sequence differences occur
between man and any single baboon species. However, most of
the sequence differences are shared among all baboon
species. A species specific nucleotide difference occurs

```
H    CTAAAGGGAAGAATAAATTAGAGAAAAATTGGAATGACTGAATCGGAACA  -250
PA                                            A T
PC                                            G T
PP                                            G T


H    AGGCAAAGGCTATAAAAAAAATTAAGCAGCACTATCCTCTTGGGGGCCCC  -200
PA
PC
PP


H    TTCCCCACACTATCTCAATGCAAATATCTGTCTGAAACGGTCCCTGGCTA  -150
PA
PC
PP

                                                _____
H    AACTCCACCCATGGGTTGGCCAGCCTTGCCTTGACCAATAGCCTTGACAA  -100
PA                        T
PC                        T
PP                        T

                   _____
H    GGCAAACTTGACCAATAGTCTTAGAGTATCCAGTGAGGCCAGGGGCCGGC   -50
PA         C            A            AG
PC         C            G            AG
PP         C            G            AG


H    GGCTGGCTAGGGATGAAGAATAAAAGGAAGCACCCTTCAGCAGTTCCACA    -1
PA   A                                  C
PC   A                                  C
PP   A                                  C
```

Figure 1. Sequence comparison of human Gγ and P.anubis
(PA), P.cynocephalus (PC), and P.papio (PP) Iγ gene promoters from -1 to -300. The duplicated CCAAT boxes are
overlined.

at -82 of Iγ gene promoter of P. anubis compared to P. cynocephalus and P. papio. The three P. anubis animals had an A,
while the two P. cynocephalus animals and P. papio animal
had a G. Humans also have a G in this position. At position
-82 of the Vγ promoter, P. anubis animals also had an A
while both P. papio and human had a G. P. cynocephalus
animals, however, had an A at this position. A G to A
difference at -265 between P. anubis and the other species

was found in only one of three P. anubis animals and, therefore, may represent a polymorphism in this species.

A comparison of HbF response following 5-azacytidine treatment with the nucleotide occurring at −82 in both the Iγ and Vγ gene promoters in clones obtained from all 6 animals is shown in Table 1.

Table 1. Maximal HbF levels following 5-azacytidine treatment in three baboon species with nucleotide sequence differences at −82 of the Iγ and Vγ genes.

BABOON	SPECIES	MAXIMAL HbF	−82 Iγ	−82 Vγ
2057	P. anubis	50.7	A	A
5005		65.5	A	A
5058		69.1	A	A
4447	P. cynocephalus	28.4	G	A
5023		34.5	G	A
3250	P. papio	21.9	G	G

All 3 P. anubis animals attain maximal HbF levels of ＞50% and have an A at −82 of both Iγ and Vγ gene promoters. The P. papio animal has the lowest maximal HbF level and has a G at −82 of both the Iγ and Vγ gene promoters. Two P. cynocephalus animals have an HbF response intermediate between the P. anubis and P. papio animals. Each P.cynocephalus contains a G at −82 of the Iγ gene promoter, but has an A at −82 of the Vγ gene promoter. The G to A difference at −82 of the γ gene promoters appears to cor- relate with the differences in maximal HbF levels attained following 5-azacytidine treatment in these three baboon species.

DISCUSSION

We have shown that a sequence difference exists at position −82 among the globin gene promoters obtained from three species of baboons. P. anubis animals have an A at −82 of both the Iγ and Vγ gene promoters, while both promoters of P. papio have a G at that position. P. cynocephalus animals

have a G at -82 of the Iγ gene promoter, but have an A in
the Vγ gene promoter. When treated with 5-azacytidine, P.
anubis animals have a high HbF response, P. papio animals
have a low HbF response, while P. cynocephalus animals
produce HbF at intermediate levels. Thus a high level of γ
gene expression is associated with the presence of A at -82.
Man, also considered to be a low HbF responder, has a G at
-82 of both fetal globin gene promoters (Shen et al 1981).
This association, however, is not proof that the molecular
basis of the species differences in HbF response is due to
the -82 difference. Therefore, we are currently testing the
effect of these base differences on γ gene promoter function
in an in vitro transcription system.

Sequence data obtained from other primates including
chimpanzee (Slightom et al 1985), orangutan (Slightom et al
1987), brown lemur (Harris et al 1986), dwarf lemur (Harris
et al 1986), gorilla (Scott et at 1984), spider monkey
(Giebel et at 1985), and galago (Tagle et at 1988) demonstrate the occurrence of a G at position -82 of the γ globin
gene promoter. The -82 G to A substitution probably origi-
nated in the baboon Vγ gene promoter and was transferred to
the Iγ gene promoter by gene conversion. In all three
baboon species, the presence of an A at -49 of the Iγ
promoter creates an Alu I site. The Vγ promoters of all
three species have a G at this position. It is likely that
the 3' boundary of the conversion event occurred 5' to this
site.

The -82 G to A change occurs 2 nucleotides 3' to the
proximal CCAAT box. The distal and proximal CCAAT boxes are
contained within 17 base pair sequences that form a pair of
direct repeats in the baboon. Figure 2 compares the baboon
-82 G to A difference with the human Aγ -117 G to A difference found in Greek HPFH (Collins et at 1985, Gelinas et
al 1985). The -117 change is located near the distal CCAAT
box, while the -82 change is located near the proximal CCAAT
sequence. Both the -117 and -82 base changes result in a G
to A substitution at opposite ends of their respective CCAAT
box sequences. From the phenotypic change associated with
these base changes, it would appear that the duplicated
CCAAT boxes function in both the repression and activation
of the γ genes. Positive and negative regulation of the sea
urchin histone H2B-1 gene occurs by binding of distinct
protein factors to the duplicated CCAAT box region of the
H2B-1 gene promoter (Barberis et al 1987). The Aγ -117

```
              -117                        -94         -82

NORMAL HUMAN  CCTTGACCAATAGCCTTgacaaggcaaaCTTGACCAATAGTCTT
GREEK HPFH    ----A-------------------------------------

P. papio/     ---------------------------C---------------
cynocephalus
P. anubis     ---------------------------C-----------A----
```

Figure 2. Comparison of nucleotide sequences of human Aγ,
Greek HPFH Aγ, P.cynocephalus Iγ, P. papio and P. anubis Iγ
from -121 to -78.

mutation could result in activation of the Aγ gene by
interfering with the binding of a repressor protein or by
increasing the affinity of an activator protein (Gumucio et
al 1988). The -82 difference results in increased gene
expression only in individuals induced to express HbF by
erythropoietic stress or drug treatment. The -82 difference
may also act by increasing the binding affinity of a
positively acting transcription factor for the CCAAT box.

REFERENCES

Barberis A, Superti-Furga G, Busslinger M (1987). Mutually
    exclusive interaction of the CCAAT-binding factor and of
    a displacement protein with overlapping sequences of a
    histone gene promoter. Cell 50:347-359.
Collins FS, Metherall JE, Yamakawa M, Pan J, Weissman SM,
    Forget BG (1985). A point mutation in the Aγ-globin gene
    promoter in Greek hereditary persistence of fetal
    hemoglobin. Nature 313:325-326.
DeSimone J, Biel SI, Heller P (1978). Stimulation of fetal
    hemoglobin synthesis in baboons, by hemolysis and hypoxia.
    Proc Natl Acad Sci USA 75:2937-2940.
DeSimone J. Heller P, Amsel J, Usman M (1980). Magnitude of
    the fetal hemoglobin response to acute hemolytic anemia
    in baboons is controlled by genetic factors. J Clin Invest
    65:224-226.
DeSimone J, Heller P, Hall L, Zwiers D (1982). 5-Azacytidine
    stimulates fetal hemoglobin synthesis in anemic baboons.
    Proc Natl Acad Sci USA 79:4428-4431.

DeSimone J, Schroeder WA, Shelton JB, Shelton JR, Espinueva Z, Huynh V, Hall L, Zwiers D. (1984). Speciation in the baboon and its relation to γ-chain heterogeneity and to the response to induction of HbF by 5-azacytidine. Blood 63:1088-1095.

Frischauf A, Lehrach H, Poustka A, Murray N (1983). Lambda replacement vectors carrying polylinker sequences. J Mol Biol 170:827-842.

Gelinas R, Endlich B, Pfeiffer C, Yagi M, Stamatoyannopoulos G (1985). G to A substitution in the distal CCAAT box of the Aγ globin gene in Greek hereditary persistence of fetal hemoglobin. Nature 313:323-325.

Giebel LB, van Santen VL, Slightom JL, Spritz RA (1985). Nucleotide sequence, evolution, and expression of the fetal globin gene of the spider monkey Ateles geoffroyi. Proc Natl Acad Sci USA 82:6985-6989.

Gumucio DL, Rood KL, Gray TL, Riordan MF, Sartor CI, Collins FS (1988), Nuclear protiens that the human γ-globin gene promoter: alterations in binding produced by point mutations associated with hereditary persistence of fetal hemoglobin. Mol Cell Biol 8:5310-5322.

Harris S, Thackeray JR, Jeffreys AJ, Weiss ML (1986) Nucleotide sequence analysis of the lemur β-globin gene family: Evidence for major rate fluctuations in globin polypeptide evolution. Mol Biol Evol 3:465-484.

Letvin NL, Linch DL, Beardsley GP, McIntyre KW, Nathan DG (1984). Augmentation of fetal hemoglobin production in anemic monkeys by hydroxurea. N Engl J Med 310:869-874.

Maxam A, Gilbert W (1980). Sequencing end-labeled DNA with base specific chemical cleavages. In Methods in Enzymology 65, New York: Academic Press, pp 499-560.

Messing J (1983). New M13 vectors for cloning. In Methods in Enzymology 101, New York: Academic Press, pp 20-78.

Papayannopoulou T, Torrealba-De Ron A, Veith A, Knitter G, Stamatoyannopoulos G (1984). Arabinosyl cytosine induces fetal hemoglobin in baboons by perturbing erythroid cell differentiation kinetics. Science 224:617-619.

Perrine RP, Pemdrey ME, John P, Perrine S, Shoup F (1978). Natural history of sickle cell anemia in Saudi Arabs. Ann Intern Med 88:1-6.

Sanger F, Nicklen S, Coulson AR (1977). Sequencing with chain terminating inhibitors. Proc Natl Acad Sci USA 74: 5463-5468.

Scott AF, Heath P. Trusko S, Boyer SH, Prass W, Goodman M, Czelusniak, Chang LYE, Slightom JL (1984). The sequence of the gorilla fetal globin genes: Evidence for multiple gene conversions in human evolution. Mol Biol Evol 1:371-389.

Shen S, Slightom JL, Smithies O (1981). A history of the human fetal globin gene duplication. Cell 26:191-203

Singer K, Chernoff AL, Singer L (1951). Studies on abnormal hemoglobins I: Their demonstration in sickle cell anemia and other hematologic disorders by means of alkali denaturation. Blood 6:413-428.

Slightom JL, Koop BF, Xu P, Goodman M (1985). Chimpanzee fetal Gγ and Aγ globin gene nucleotide sequences provide further evidence of gene conversions in hominid evolution. Mol Biol Evol 2:370-389.

Slightom JL, Thiesen TW, Koop BF, Goodman M (1987). Orangutan fetal globin genes: Nucleotide sequences reveal multiple gene conversions during hominid evolution. J Biol Chem 262:7472-7483.

Tagle DA, Koop BF, Goodman M, Slightom JL, Hess DL, Jones RT (1988). Embryonic and γ globin genes of a prosimian primate (Galago Crassicaudatus): Nucleotide and amino acid sequences, developmental regulation and phylogenetic footprints. J Mol Biol 203:439-455.

Hemoglobin Switching, Part A: Transcriptional Regulation, pages 269–277

NEGATIVE REGULATION OF THE HUMAN EMBRYONIC GLOBIN GENES ζ AND ε

P. Lamb, P. Watt and N.J. Proudfoot.

Sir William Dunn School of Pathology, University of Oxford, South Parks Rd., Oxford OX1 3RE.

INTRODUCTION

Mutational analysis of human globin genes has revealed the presence of a variety of transcriptional control elements involved in the regulation of these genes both in tissue culture cells and in transgenic mice (e.g. Antoniou et al. 1988; Anagnou et al. 1986; Grosveld et al. 1987; Myers et al. 1986). It is likely that these sequences and the factors that interact with them are intimately involved in the switches in globin gene expression that occur during development. To date, most work has concentrated on the human adult and foetal genes α,ß and γ. We have been studying sequences involved in the control of expression of the embryonic genes, ζ and ε.

In humans, the globin polypeptides encoded by the embryonic genes are synthesised in primitive nucleated erythroblasts produced by the blood islands of the yolk sac. Synthesis is thought to be initiated when erythropoeisis begins at about the fourth week of embryogenesis, and continues until the fifth week, when ζ and ε production drops abruptly. By the sixth week, little ζ globin is detected, whereas ε globin is detectable until the seventh week (Peschle et al. 1985). In mice, the levels of the different globin proteins found in the peripheral blood during development is closely paralleled by the respective globin mRNA levels, indicating that control of transcription and/or message stability determines protein levels (Whitelaw et al. 1989). Previous studies have shown that the cloned human ζ globin gene is active when introduced into K562 cells (early erythroid), but not when transfected into HeLa or COS cells (Proudfoot et al. 1984; Norman et al. 1987). We have analysed the promoter region of the human ζ gene in order to identify regulatory sequences that are involved in the control of transcription of the gene. In addition, we have initiated studies on the promoter of the human ε globin gene, to allow a comparison of the promoters of these coordinately regulated genes.

RESULTS

Deletion Analysis of the ζ Globin Promoter.

In order to study the ζ promoter in isolation, we fused it to the reporter gene CAT. Initially, 1.1Kb of ζ 5' flanking sequence, up to and including the cap site was linked to the CAT gene (Figure 1).

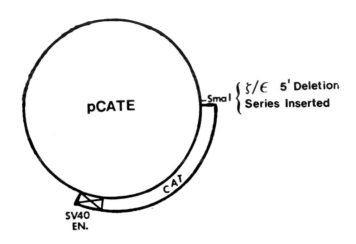

Figure 1. Structure of pCATE.
The CAT expression vector pCATE contains the CAT gene,SV40 small t intron and early polyadenylation signal and the SV40 enhancer (hatched box). Promoter fragments are inserted into the unique Sma1 site 5' of the CAT gene.

A series of 5' deletion mutants was then generated using BAL 31 or restriction enzyme sites. These plasmids were introduced into Putko cells, together with a control plasmid that expresses ß-galactosidase. Putko cells are a derivative of K562 cells that maintain their early erythroid characteristics but which transfect with greater efficiency. After 48 hours, cell lysates corresponding to equal amounts of ß-galactosidase activity were used to perform CAT assays, and the percentage conversion determined. The averaged results of 6 experiments are shown in Figure 2. Constructs with 147 bp or more of ζ 5' flanking sequence give levels of CAT activity 3-4 times background. However, deletion to -137 (relative to the cap site, 0) produces a 3-fold increase in CAT activity suggesting the presence of an inhibitory sequence in this region. Further deletion suggests that a second inhibitory sequence may be present at -115, removal of which also gives a 3-fold increase in CAT activity. These effects are not dependent on the presence of the SV40 enhancer; deletion from -550 to -113 results in a

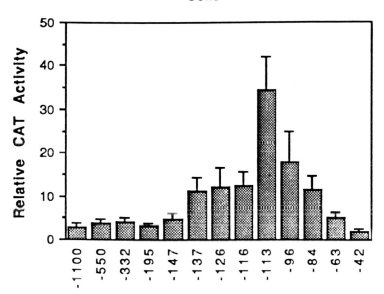

Figure 2. Relative Activity of the ζ-CAT Deletion Series in Putko Cells.
The CAT activities of the ζ promoter 5' deletion mutants are shown relative to the promoterless construct pCATE (whose activity is designated as 1). The average of 6 experiments is shown for each construct. The vertical extension to each column represents the standard error in the mean (s.e.m.).

10-fold increase in CAT activity even in an enhancerless plasmid (data not shown). Beyond -113, deletions produce stepwise decreases in CAT activity, consistent with the removal of positively acting elements. These deletions remove the sequences GATAAG, CCACCC and CCAAT that have been identified as being functionally important in other globin gene promoters and enhancers (Emerson et al. 1987; Evans et al. 1988; Myers et al. 1986). The deletion analysis thus identifies several putative regulatory elements in the ζ promoter; two negative sequences, together with positive control elements that are found in other globin genes.

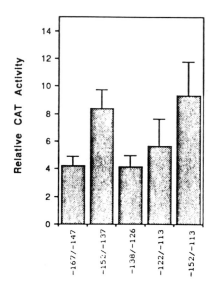

**Activity of Zeta LS Mutants
In Putko Cells**

```
            -170        -160        -150        -140        -130        -120        -110
            GGAACAGGAGTGACAGCCCCCAAACCCCAGTCCCACAGCCCTGAGGGCCCCTTTGTCACTGGA    WT

            -----        cgAG CtCggtA ------------------------------------------    -167/-147

            ----------------------- gAgCtCggtaCCCc-----------------------------    -152/-137

            --------------------------------- gAGCtC Ggtac----------------------    -138/-126
                                                                ag        c
            ----------------------------------------------------gcTcGgtACc---    -122/-113

            -------------------          gagctcggtacc                    ---    -152/-113
```

Figure 3. Relative Activity of the ζ promoter Linker Scanning Mutants in Putko Cells.

The CAT activities of the linker scanning mutants in the -167 to -113 region are shown relative that of pCATE. The average of 6 experiments is presented with the s.e.m. indicated by the vertical line. The sequence of each of the mutants is given underneath the graph, compared to the wildtype sequence (WT). The solid lines represent unmutated regions. Linker sequences are shown by lower case type except where they fortuituosly match the ζ sequence in which case upper case type is used. Blank spaces indicate deleted bases whereas letters above the line are inserted bases.

Linker Scanning Mutants in the -167 to -113 Region.

We constructed a set of linker scanning mutants designed to further characterise the two putative negative elements. A 14 bp linker was introduced at various positions between -167 and -113 within the promoter (Figure 3). In addition, an internal deletion mutant was created that should remove both negative elements. These constructs were transfected into Putko cells as before and the relative CAT activities obtained are shown in Figure 3. Mutants -167/-147 and -138/-126 give wild type levels of activity. However mutant -152/-137 gives about 2-fold higher levels of activity, in good agreement with the deletion analysis and confirming the presence of a negative element at about -140. Surprisingly, mutant -122/-113 gives essentially wild type activity and the internal deletion gives increased activity consistent only with the removal of the -140 element. The negative sequence suggested by the deletion analysis at -115 is not evident in these assays. It is possible that sequences upstream may modulate the activity of this element such that it is only seen when these sequences are removed. Further clustered point mutants are being constructed in a deleted promoter to test this point.

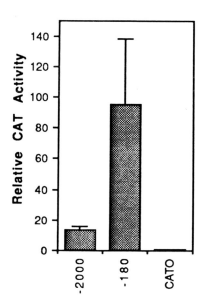

Activity of Epsilon/CAT Constructs in Putko Cells

Figure 4. Relative Activity of the ε-CAT Constructs in Putko Cells. The CAT activity of 2 ε promoter-CAT constructs is shown relative to the promoter and enhancer minus plasmid pCATO. The average of 6 experiments is shown together with the s.e.m. (vertical line).

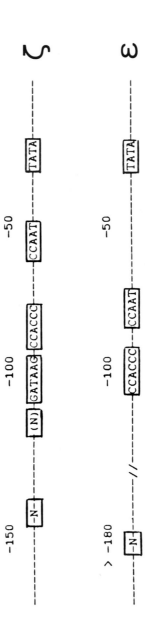

Negative Regulation of the ε Globin Promoter.

We have started an analysis of the human ε globin promoter using the methodology outlined above. An ε-CAT fusion gene containing 2.0 Kb of 5' flanking sequence, and a deleted construct have been expressed in Putko cells. The levels of CAT activity obtained are shown in Figure 4. Removal of sequences between -2.0 Kb and -0.18 Kb results in a 10-fold increase in activity, indicating that the ε promoter is also negatively regulated. We are currently defining the inhibitory sequences further.

DISCUSSION.

This mutational study of the human ζ globin promoter shows that it includes both positive and negative acting sequences. The presence of a negative element is particularly interesting since it may be involved in the reduction of ζ gene expression that occurs after the fifth week of development. We predict that factor(s) which bind to this sequence will be absent or inactive in early primitive erythroblasts, but will be active at later stages. Putko cells may represent an intermediate stage. They are capable of expressing their ζ genes, but not at maximal levels. Identification and characterisation of factors that interact with this element is clearly important. The finding that the ε globin promoter is also under negative control is provocative given the similar regulation patterns of the ζ and ε genes *in vivo*. Figure 5 shows a comparison of the ζ and ε globin promoters.

As well as the positive acting control sequences found in most globin gene promoters, both possess upstream negative control element(s). The relationship between these negative elements in the ζ and ε promoters is at present unknown. However,an exciting possibility is that a common factor or factors is responsible for repressing expres-sion of both human embryonic globin genes in a developmentally regulated fashion.

Figure 5. A Comparison of ζ and ε Promoter Organisation.
Positive acting control sequences identified by comparison with other globin genes are shown boxed. The negative regulatory elements identified in this study are indicated by an N. The numbers above each line are the distance in nucleotides from the respective cap sites (designated 0). The slash marks in the ε promoter line show a change in scale.

MATERIALS AND METHODS.

Plasmid Constructions.

The CAT expression plasmid pCATO contains the CAT gene, SV40 intron and poly A site from the plasmid SV0CAT (Gorman et al. 1982) cloned into the BamH1 site of PUC 119. A 200 bp fragment containing the SV40 enhancer was then blunt ended into the reformed BamH1 site at the 3' side of the CAT gene to give plasmid pCATE. A 1.1 Kb Sty1 fragment (-1.1 Kb to +4 bp) and an EcoR1-Sty1 fragment (-550 bp to +4 bp) from the promoter of the human ζ globin gene (Proudfoot et al. 1982) were then blunt ended and inserted into the Sma1 site immediately upstream of the CAT gene. BAL 31 deletions were generated by linearising the -550 construct with Sac1, digesting with BAL 31 for various times, recutting with HindIII and isolating the shortened ζ promoter-CAT fragments from an agarose gel. These were ligated into HindIII/Sma1 cleaved PUC 119. The 3' deletions were similarly generated by BAL 31 digestion from the Pst1 site (-84) in a ζ promoter extending to -550. A Sac1 linker was ligated to the deleted end, and matching 5' and 3' deletions joined via the Sac1 site. All deletions were characterised by Maxam and Gilbert sequencing.

The ε-CAT constructs were generated by ligating blunt ended EcoR1-PvuII (-2.0 Kb to +12 bp) or BamH1-PvuII (-179 bp to +12 bp) fragments from the 5' flanking region of the human ε globin gene (Baralle et al. 1980) into Sma1 cut pCATO.

The ß-galactosidase control plasmid contains the lac z gene under the control of a ß-actin promoter.

Cell Culture, Transfection and Enzymatic Assays.

Putko cells, a hybrid between K562 cells and a B-cell lymphoma (Klein et al. 1980) were grown in RPMI containing 10% foetal calf serum, 2 mM glutamine, 100µg/ml penicillin and 100U/ml streptomycin. Cells ($5x10^6$-10^7) were transfected by electroporation in standard Hepes buffered saline containing 0.5% PEG 6000, 50µg test plasmid and 5µg of control plasmid. A shock of 1750 V was delivered using a Biorad GenePulsar set at 25µF capacitance. About 48 hours later the cells were harvested, lysed by sonication, and the cleared lysates quantitated for ß-galactosidase activity (Herbomel et al. 1984). Lysates corresponding to equal amounts of ß-galactosidase activity were heated to 65°C for 5 minutes then used in a standard CAT assay (Gorman et al. 1982). The percentage of chloramphenicol converted was quantitated by cutting the chloramphenicol and 3-acetyl chloramphenicol from the plates and counting them in a liquid scintillation counter.

REFERENCES.

Anagnou N.P., Karlsson S., Moulton A.D., Keller G. and Nienhuis A.W. (1986) EMBO J. 5 121-126

Antoniou M., deBoer E., Habets G. and Grosveld F. (1988) EMBO J. 7 377-384

Baralle F., Shoulders C.C. and Proudfoot N.J. (1980) Cell 21 621-626

Emerson B.M., Nickol J.M., Jackson P.D. and Felsenfeld G. (1987) Proc.Natl.Acad.Sci. 84 4786-4790

Evans T., Reitman M. and Felsenfeld G. (1988) Proc.Natl.Acad.Sci. 85 5976-5980

Gorman C.M., Moffat L.F. and Howard B.H. (1982) Mol.Cell Biol. 2 1044-1051

Grosveld F., Blom van Assendelft G., Greaves D. and Kollias G. (1987) Cell 51 975-985

Herbomel P., Bourachot B. and Yaniv M. (1984) Cell 39 653-662

Klein G., Zeuthen J., Eriksson I., Tekasaki P., Bernoco M., Rosen A., Masucci G., Povey S. and Ber R. (1980) J.Natl.Cancer Inst. 64 725-738

Kollias G., Wrighton N., Hurst J. and Grosveld F. (1986) Cell 46 89-94

Myers R.M., Tilly K. and Maniatis T. (1986) Science 232 613-618

Norman C., Raymond V. and Proudfoot N.J. (1987) In:Developmental Control of Globin Gene Expression, Prog.in Clinical and Biological Research 251 235-252

Peschle C., Malvilio F., Care A., Migliaccio G., Migliaccio A.R. Salvo G., Samoggia P., Petti S., Guerriero R., Marinucci M., Lazzaro D., Russo G. and Mastroberardino G. (1985) Nature 313 235-238

Proudfoot N.J., Gil A. and Maniatis T. (1982) Cell 31 553-563

Proudfoot N.J., Rutherford T.R. and Partington G.A. (1984) EMBO J. 3 1533-1540

Whitelaw E., Lamb P., Hogben P. and Proudfoot N.J. (1989) this Volume

Hemoglobin Switching, Part A: Transcriptional Regulation, pages 279–289
© 1989 Alan R. Liss, Inc.

NEGATIVE CONTROL OF THE HUMAN ε-GLOBIN GENE

Shi Xian Cao, Pablo D. Gutman, Harish P.G. Dave
and Alan N. Schechter

Laboratory of Chemical Biology, National
Institute of Diabetes, and Digestive and Kidney
Diseases, National Institutes of Health,
Bethesda, MD 20892

INTRODUCTION:

The human ε-globin gene is a member of the β-like globin gene family. It is expressed in a developmental and tissue-specific manner, i.e. at the embryonic stage and in the yolk sac (see Karlsson and Nienhuis, 1985 for review). At the end of the embryonic stage, the ε-globin gene is switched off and the $^G\gamma$ and $^A\gamma$ genes are activated. At the present time, the mechanisms governing this switch are not known.

In an attempt to understand the mechanisms controlling the expression of the human ε-globin gene, we have studied the 5' flanking sequences of the gene. By introducing a series of 5' deletion mutants into cultured cells, we have identified a DNA fragment between -392 and -177 relative to the cap site that negatively regulates ε-globin promoter activity. This negative control element acts both 5' and 3' of the transcription unit, and in an orientation independent manner, thus complying with the definition of a silencer (Brand et al., 1985). Comparison of the inhibitory effect in HeLa cells and K562 cells showed that the negative effect of the silencer is stronger in the former. Sequence analysis showed that two short regions within the silencer element share extensive homology with each of two silencer sequences that have

been identified in the chicken lysozyme and rat insulin 1
genes, as well as with negative control sequences found
in several other genes (Baniahnad et al. 1987).

Our results suggest that this silencer element may
play an important role in the regulation of the human
ε-globin gene activity in a tissue specific- or
developmental stage-specific manner.

RESULTS

A Negative Control Element Exists in the 5' Region of the
ε-Globin Gene

To identify DNA sequences that have positive or
negative control function, we have constructed a series of
5' deletion mutants fused to the bacterial chloramphenicol
acetyl transferase (CAT) gene (Fig. 1A). These plasmids
were introduced into either the human erythroid leukemia
cell line K562, or HeLa cells. Analysis of CAT activities
reveal a similar expression pattern in both K562 cells and
HeLa cells: pεΔ392 shows the lowest CAT level and pεΔ177
shows the highest CAT level among all the deletion mutants
(data not shown). A comparison of the CAT assays between
pεΔ392 and pεΔ177 is shown in Fig. 2. This result suggests
that the DNA sequence between -177 and -392 is a negative
control element for the ε-globin promoter. To quantitate
the differences in CAT activity betweeen pεΔ177 and
pεΔ392, we normalized the CAT level of pεΔ392 to pεΔ177
and the results are shown in Table 1. While the negative
control element inhibits ε-globin promoter activity in
both cell lines, the level of inhibition is higher in HeLa
cells (10 fold) than in K562 cells (3 fold).

Table 1. CAT Activities of 5' Deletion
Mutants in HeLa Cells and K562 Cells

	pεΔ177	pεΔ392
HeLa	1.00 ± 0.07	0.11 ± 0.02
K562	1.00 ± 0.04	0.37 ± 0.03

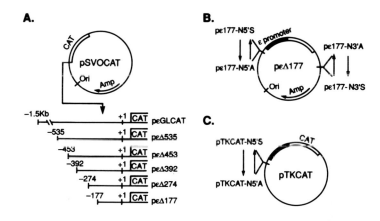

Figure 1. Plasmid structures. A. pεGLCAT and 5'
deletion mutants. Numbers indicate the length of the 5'
flanking sequences relative to cap site. B. Plasmids
pε177-N5'S, pε177-N5'A, pε177-N3'S and pε177-N3'A. The
215 bp negative control region was subcloned into pεΔ177
either 5' or 3' of the transcriptional unit in both sense
and antisense orientations to generate the four plasmids.
C. Plasmids pTKCAT-N5'S and pTKCAT-N5'A. The 215 bp
negative control region was inserted 5' of the TK promoter
in either sense or antisense orientation.

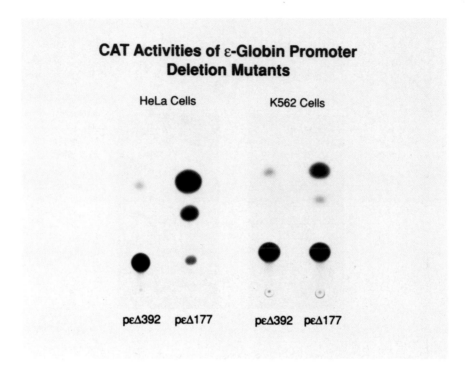

Figure 2. CAT activities of pεΔ177 and pεΔ392 in HeLa cells and K562 cells. HeLa cells (1x10⁶) or K562 cells (5x10⁶) were plated on a 100mm tissue culture dish. After 24 hours incubation, each dish was transfected with 10μg of plasmid and incubated for 48 hours. Cells were lysed and equal amounts of protein (500μg for HeLa cells and 1000μg for K562 cells) was used for CAT assay.

RNA analysis showed that the CAT mRNA was correctly
initiated and the level of CAT mRNA from cells transfected
with pεΔ177 was much higher than that transfected with
pεΔ392 (Fig. 3). This result is consistent with the CAT
assay results and indicates that the inhibition is exerted
on the RNA level.

The Negative Control Region Functions as a Silencer

In many cases, a negative control region can function
in a position- and orientation-independent manner
(Baniahmad et al. 1987). We next examined whether the
ε-globin gene negative control region can act in this
manner. Fig. 1B shows the constructions used in this
study. The 215 bp fragment was cloned into either 5' or
3' of the transcriptional unit and in both orientations to
generate the plasmids pε177-N5'S, pε177-N5'A, pε177-N3'S
and pε177-N3'A respectively. These plasmids were
introduced into HeLa cells and K562 cells. The CAT assay
results are shown in Table 2. First, the negative
regulatory effect of the fragment can be observed when it
is placed both 5' and 3' of the transcriptional unit and
in both orientations, even though the effect is weaker
when it is located 3'. This suggests that the ε negative
control region is a silencer. Second, the inhibitory
effect is stronger in HeLa cells than in K562 cells. For
example, when the negative control region was placed 5' of
the ε promoter, it produced 14 and 25 fold inhibition for
sense and antisense orientation in HeLa cells compared
with 2-3 fold inhibition in K562 cells.

Table 2. Effects of the Negative Control Region
on the Expression of pεΔ177

	pεΔ177	pε177-N5'S	pε177-N5'A	pε177-N3'S	pε177-N3'A
HeLa	1.00±0.14	0.07±0.02	0.04±0.01	0.12±0.04	0.27±0.05
K562	1.00±0.08	0.35+0.04	0.36±0.06	0.43±0.04	0.45±0.11

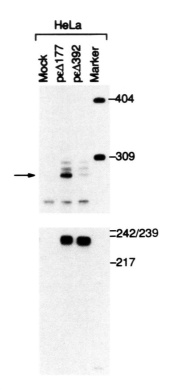

Figure 3. RNA protection analysis of HeLa cells transfected with pεΔ177 and pεΔ392. HeLa cells were transfected with either pεΔ177 plus pSV2Neo or pεΔ392 plus pSV2Neo. Total RNA was isolated and analyzed by use of a riboprobe in RNAse T1 mapping as described by Promega Biotec. Arrow indicates the band protected by correctly initiated CAT mRNA. The bands on the lower panel are internal control generated by Neo mRNA. The upper panel was exposed for 25 hours and the lower panel was exposed for 4 hours.

The Silencer Inhibits the Expression From a Heterologous
Promoter

Next the ε silencer was tested for its effect on a
heterologous promoter. The 215 bp fragment was inserted
5' of the TKCAT transcriptional unit in both orientations
(Fig. 1C). The resulting plasmids were transfected into
HeLa cells and CAT activity was measured. Table 3 shows
that the activity of the TK promoter is also inhibited by
the ε silencer.

Table 3. Effects of the Negative Control
Region on the Expression of pTKCAT

pTKCAT	pTKCAT-N5'S	pTKCAT-N5'A
1.00 ± 0.04	0.25 ± 0.12	0.10 ± 0.04

Sequence Comparisons

Recently Baniahmad et al. (1987) have compared the
DNA sequences of negative regulatory elements from a
number of genes including chicken lysozyme, rat insulin 1,
human β-interferon, mouse IgH, rat growth hormone and
mouse major histocompatibility class I gene. They
reported two short sequences (termed box 1 and box 2) in
these negative regulatory elements that shared extensive
sequence homology among these genes (Baniahmad et al.
1987). We therefore compared the ε silencer with these
sequences. We found that the two homology boxes also
exist in the ε-silencer element, but in reversed
orientations (Fig. 4). The sequence of region 1 in the ε
silencer matches 8 out of 9 nucleotides to the box 1
consensus sequence (Fig. 4B). The sequence of region 2 in
ε silencer matches 8 out of 10 nucleotides to the box 2
consensus sequence (Fig. 4C).

A.

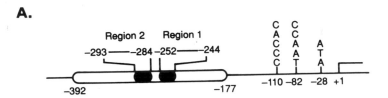

B. **Homology Region 1**

ANCCTCTC$_C^T$ Consensus
ACCCTCT t C   Human ε-Globin

C. **Homology Region 2**

ANTCTCCTCC   Consensus
ATTCTCCT t t   Human ε-Globin

Figure 4.   Comparison of the ε-silencer and the negative consensus sequences.   A.   Position of the ε-silencer and the two homologous regions in the ε-globin gene 5' flanking region.   B.   Homology region 1 and the consensus sequence of box 1.   C.   Homology region 2 and the consensus sequence of box 2.   The two consensus sequences of box 1 and box 2 are derived from Baniahmad et al. 1987.

DISCUSSION

We have identified a transcriptional silencer which negatively regulates the expression from the ε-globin gene promoter. The silencer element is located between -177 and -392 bp relative to the cap site. This region has been shown to contain a DNase I hypersensitive site at -200 bp in K562 cells (Tuan et al. 1984); therefore it is in an active chromatin conformation. Considering the close proximity of the silencer to the promoter elements, we believe that the ε silencer is very likely to participate in the regulation of ε-globin gene activity.

DNA elements having negative regulatory function have been recorded in many systems. Some of them play important roles in regulation of tissue specific gene expression (Kadesch et al. 1986, Larsen et al. 1986, Muglia et al. 1986, Nir et al. 1986, Colantuoni et al. 1987). Some are involved in gene induction or repression in response to hormonal stimulus (Gaub et al. 1985, Osborne et al. 1985, Charron and Drouin, 1986, Goodbourn et al. 1986). For some negative regulatory elements, no specific physiological functions have been defined (Rosen et al. 1985, Laimins et al. 1986, Remmers et al. 1986). For the ε-globin gene silencer, we speculate that it could act either in a tissue specific manner or in a developmental stage specific manner. For example, the silencer could be active in non-erythroid tissues or at the end of the embryonic stage during development; and therefore shut down ε-globin expression. The fact that the silencer has a stronger effect in HeLa cells than in K562 cells supports the idea that this silencer plays a role in conferring tissue specific expression to the ε-globin gene. On the other hand, the fact that it is partially active in K562 cells (which seem to be arrested in an embryonic-fetal stage) may indicate it acts during the developmental switching process.

ACKNOWLEDGEMENTS

We thank Mrs. Laura Barry for preparation of this manuscript.

REFERENCES

Baniahmad A, Muller M, Steiner C, Renkawitz R (1987).
Activity of two different silencer elements of the
chicken lysozyme gene can be compensated by enhancer
elements. EMBO J 6:2297-2303.

Brand AH, Breeden L, Abraham J, Sternglanz R, Nasmyth K
(1985). Characterization of a "silencer" in yeast: a
DNA sequence with properties opposite to those of a
transcriptional enhancer. Cell 41:41-48.

Charron J, Drouin J (1986). Glucocorticoid inhibition of
transcription from episomal proopiomelanocortin gene
promoter. Proc Natl Acad Sci USA 83:8903-8907.

Colantuomi V, Pirozzi A, Blance C, Cortese R (1987).
Negative control of liver-specific gene expression:
cloned human retinal-binding protein gene is repressed
in HeLa cells. EMBO J 6:631-636.

Gaub MP, Dierich A, Astinotti D, LePennec JP, Chambon P
(1985). In Gluzman Y (ed.) Current Communications in
Molecular Biology. Eukaryotic Transcription. Cold
Spring Harbor Laboratory Press. New York, pp. 123-131.

Goodbourn S, Burstein H, Maniatis T (1986). The human
β-interferon gene enhancer is under negative control.
Cell 45:601-610.

Kadesch T, Zerros P, Ruezinsky D (1986). Functional
analysis of the murine IgH enhancer: evidence for
negative control of cell-type specificity. Nucleic
Acids Res 14:8209-8221.

Karlsson S, Nienhuis AW (1985). Developmental regulation
of human globin genes. Ann Rev Biochem 54:1071-1108.

Laimins L, Holmgren-Konig M, Khoury G (1986).
Transcriptional "silencer" element in rat repetitive
sequences associated with the rat insulin 1 gene locus.
Proc Natl Acad Sci USA 83:3151-3155.

Larsen PR, Harney JW, Moore DD (1986). Repression
mediates cell-type-specific expression of the rat growth
hormone gene. Proc Natl Acad Sci USA 83:8283-8287.

Muglia L, Rothman-Denes LB (1986). Cell type-specific
negative regulatory element in the control region of the
rat α-fetoprotein gene. Proc Natl Acad Sci USA
83:7653-7657.

Nir U, Walker MD, Rutter WJ (1986). Regulation of rat
insulin 1 gene expression: evidence for negative
regulation in nonpancreatic cells. Proc Natl Acad Sci
USA 83:3180-3184.

Osborne TF, Goldstein JL, Brown MS (1985). 5' End of HMG CoA reductase gene contains sequences responsible for cholesterol-mediated inhibition of transcription. Cell 42:203-212.

Remmers EF, Yang JQ, Marcu KB (1986). A negative transcriptional control element located upstream of the murine c-myc gene. EMBO J 5:899-904.

Rosen CA, Sodroski JG, Haseltine WA (1985). The location of cis-acting regulatory sequences in the human T cell lymphotropic virus type III long terminal repeat. Cell 41:813-823.

Tuan D, London IM (1984). Mapping of DNaseI-hypersensitive sites in the upstream DNA of human embryonic ε-globin gene in K562 leukemia cells. Proc Natl Acad Sci USA 81:2718-2722.

Hemoglobin Switching, Part A: Transcriptional Regulation, pages 291–299

A FACTOR FROM ERYTHROLEUKEMIA-LIKE K562 CELL NUCLEI BINDS
INTRON A OF EMBRYONIC AND FETAL GLOBIN GENES

M.L. Brown, G.K.T. Chan, L. Locklear, N. Kawamura
and D.P. Bazett-Jones

Department of Medical Biochemistry, University of
Calgary, Calgary, Alberta T2N 4N1

INTRODUCTION

Elaboration of the control of gene expression at the
molecular level remains a fundamental problem in modern cell
biology. After the onset of zygotic expression in the
developing embryo, the control of gene expression lies
primarily at the level of transcription initiation.
Important questions remain to be answered in relation to the
differential transcription of multi-gene families, such as
beta-globins, during development.

The beta-globin gene complex occupies 60 Kb of DNA in
the short arm of chromosome 11 (11p15.5). This complex
provides a particularly attractive model system for studying
gene regulation in that globin expression is restricted to
a single cell lineage and the genes of the complex are
sequentially activated and repressed in a tightly controlled
fashion during development (Muller et al., 1988). Several
studies including altered globin expression in transient
chicken-human heterokaryons (cell fusion without nuclear
fusion) and in vitro transcription studies (see Lanfranchi
et al., 1984 and Bazett-Jones et al., 1985) suggest that
trans-acting protein factors are involved in the tissue
specific, differential expression of globin genes during
development.

Many approaches have been utilized to delineate the
factors involved in the tissue specific and sequential
transcription of globin genes. Protein binding sites within

and flanking the globin genes have been physically characterized by gel retardation assays (Hendrickson, 1985) and several DNA "footprinting" techniques such as DNase 1 protection (Galas and Schmitz, 1978). in vitro transcription experiments (Bazett-Jones et al., 1985; Wada and Noguchi, 1988) along with several reverse genetic techniques utilizing transient and stably transfected cell lines as well as transgenic mice have provided insights into the function of these DNA binding factors (Atweh et al., 1988; Miller et al., 1988). We are currently utilizing a novel approach to study specific globin DNA-protein interactions using electron microscopy.

Electron spectroscopic imaging (ESI) allows for the direct visualization of the ultrastructure and elemental composition of biological specimens (Bazett-Jones and Ottensmeyer, 1981). This technique has been applied to the study of DNA-protein interactions (Bazett-Jones et al., 1988), RNA-protein complexes (Bazett-Jones, 1988) as well as the secondary structure of specific DNA-transcription factor complexes (Bazett-Jones and Brown, 1989). In this report we describe the demonstration of an intron-specific binding factor and its preliminary characterization by ESI.

MATERIALS AND METHODS

Gel mobility shift assays were performed by pre-incubating non-specific competitor nucleotides (600 ug/mL yeast tRNA, 40 ug/mL pBR322 plasmid and 60 ug/mL mixed penta nucleotides) with crude K562 nuclear extract (see Bazett-Jones et al., 1985) in binding buffer (25 mM HEPES(pH 7.5), 25 mM NaCl, 2 mM $NaPO_4$, 5 mM $MgCl_2$, 0.1 mM EGTA, 0.5 mM DTT and 5 % glycerol) for 5 minutes at room temperature. 5′ end-labelled DNA fragments of interest, were then added and incubated for a further 30 minutes. Resulting protein-DNA complexes were then subjected to gel electrophoresis (see Hendrickson, 1985).

Protein-DNA complexes for ESI analysis were prepared by incubating 10 ug of linearized DNA with 12 ug total crude K562 nuclear protein in 60 uL binding buffer as above, except that 50 ug yeast tRNA and 25 ug Calf thymus DNA immobilized onto cellulose were used as non-specific competitors. Protein-DNA complexes were then purified by gel exclusion chromatography (Biogel A5M). Purified protein-DNA complexes were diluted several fold with 10 mM Tris/5 mM MgCl₂ and dried onto 2-3 nM carbon support films covering 1000 mesh electron microscope grids.

Electron spectroscopic imaging was performed, as earlier described (Bazett-Jones et al., 1988), on a Zeiss EM 902 electron microscope equipped with a prism-mirror-prism type electron imaging spectrometer. The instrument was operated with an accelerating voltage of 80 keV. Reference images and phosphorous enhanced images were recorde at 120 and 170 eV energy loss respectively (see Bazett-Jones et al., 1988).

RESULTS AND DISCUSSION

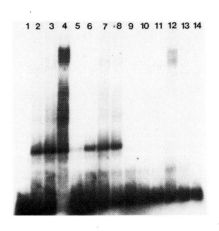

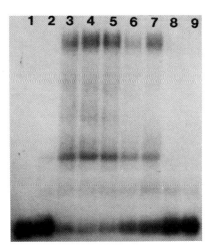

Figure 1 Gel Mobility Shift
Self Competition

Figure 2 Gel Mobility Shift
Epsilon Competition

Gel retardation studies (data not shown) of G-gamma globin 5' flanking regions incubated with induced K562 extracts, demonstrated the binding of multiple factors to promoter regions that have been characterized previously (see Superti-Furga et al., 1988 and Mantovani et al., 1988). Fragments containing the 123 bp intron A of G-gamma globin also demonstrated gel retardation as illustrated in Figure 1. 500 picograms of the 123 bp BstEII-AccI (+155-+278) fragment, containing most of the human G-gamma intron A, was incubated with increasing amounts of induced K562 nuclear extract (0, 0.5, 0.75 and 1.0 ug in Lanes 1-4 respectively). In Lanes 5 and 6, 0.5 ug of crude extract were pre-incubated with an additional 50 and 500 ng of vector DNA as non-specific competitor before addition of the labelled fragment. Lanes 7 and 8 are identical to lanes 5 and 6 except that 0.75 ug of extract was used. In Lanes 9 to 11, 0.5 ug of K562 extract was pre-incubated with cold fragment in 100, 500 and 1,000 fold excess. In lanes 12 to 14, 0.75 ug of K562 nuclear extract was incubated with cold competitor again in 100, 500 and 1,000 fold excess. These results indicate that a factor or factors in induced K562 nuclear extract binds with specificity to the first intervening sequence of G-gamma globin.

EPSILON GLOBIN INTRON A
GTAAGCATTGGTTCTCAATGCATGGGAATGAAGGGTGAATATTACCCTAGCAAGTTGATTG
GGAAAGTCCTCAAGATTTTTTGCATCTCTAATTTTGTATCTGATATGGTGTCATTTCATAG

^GGAMMA-GLOBIN INTRON A
GTAGGCTCTGGTGACCAGGACAAGGGAGGGAAGGAAGGACCCTGTGCCTGGCAAAAGTCCAG
GTCGCTTCTCAGGATTTGTGGCACCTTCTGACTGTCAAACTGTTCTTGTCAATCTCACAGG

BETA-GLOBIN INTRON A
GTAGGCTCTGGTGACCAGGACAAGGGAGGGAAGGAAGGACCCTGTGCCTGGCAAAAGTCCAG
GTCGCTTCTCAGGATTTGTGGCACCTTCTGACTGTCAAACTGTTCTTGTCAATCTCACAGG

Figure 3 Globin Intron A Sequences

Exonuclease III footprint analysis (manuscript in preparation) maps this binding activity to a sequence, AAAGTCC also found within the epsilon intron A but not in the beta intron A as illustrated in Figure 3. Cold DNA fragments containing the first intervening sequence of epsilon globin are able to compete off this binding activity as shown in Figure 2. 500 picograms of labelled fragment as in Figure 1 was incubated with 0, 0.5 and 1.0 ug of K562 nuclear extract (Lanes 1 to 3). In lanes 4 to 6; 1, 10, and 100 fold mass excess of vector DNA was pre-incubated with 1.0 ug of K562 extract before probe addition. In Lanes 7 to 9, the 431 bp NcoI fragment of epsilon globin gene containing intron A was pre-incubated with 1.0 ug of extract in 1, 10 and 100 fold mass excess before addition of labelled probe. Thus while non-specific DNA is unable to compete for gamma globin intron A binding activity, the intron A of epsilon globin which also contains the sequence AAAGTCC is able to compete for binding factor(s). This sequence is interesting in that it occurs in several transcriptionally relevant regions including the 72 bp repeat of SV40 virus, the LTR of murine leukaemia virus, the highly conserved promoter of prolactin genes, as well as the enhancers of lymphotropic papova virus,

murine amylase and immunoglobulin heavy chain genes
(Wingender, 1988 and Gutierrez-Hartman, 1987). Recently, the
first intron of the alpha1(I) collagen gene has proven to be
important for transcription (Bornstein et al., 1988 and
Killen et al., 1988). The first intervening sequence of
globin genes may prove to have a role in transcription as has
been suggested for the larger second intron (Galson et al.,
1988).

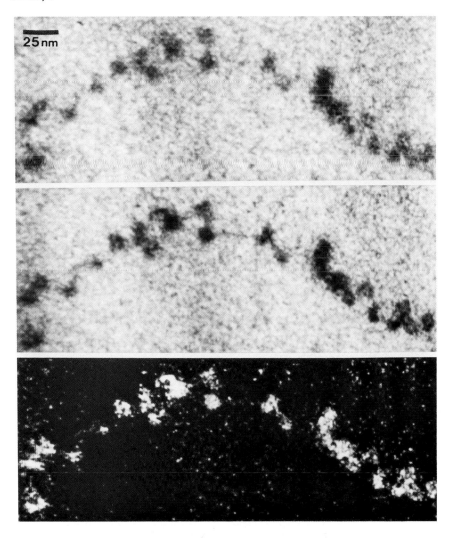

Figure 4 ESI Analysis of Yeast Chromatin

Globin DNA-protein complexes are currently being studied using ESI. To establish the validity of this method as a tool for studying protein-DNA interactions, a variety of structures that have been well-defined by biochemical methods have been analyzed (Bazett-Jones et al.,1988). These include the ultrastructure of the RNA polymerase III transcription factor TFIIIA bound to its template 5SDNA (Bazett-Jones and Brown, 1989). Figure 4 illustrates ESI analysis of <u>Saccharomyces</u> <u>cerevisiae</u> chromatin. The reference image in the top panel was taken at an energy loss of 120 eV, before the phosphorous ionization edge. The phosphorous-enhanced image (middle panel) taken at an energy loss of 170 eV, on the phosphorous ionization peak, shows increased contrast of the DNA as electrons undergoing inelastic collisions with phosphorous are selectively imaged. The bottom panel represents the net phosphorous signal after the subtraction of the reference image from the phosphorous enhanced image. The phosphate molecules of the DNA can be seen on the periphery of some of the nucleosome cores as expected from the biochemical data.

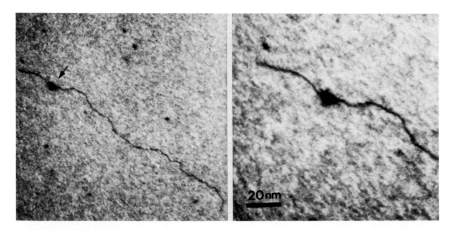

Figure 5 Phosphorous Enhanced Image of Gamma Intron A-Protein Complex

The phosphorous enhanced image of a K562 protein bound to intron A DNA of the gamma globin gene is shown in Figure 5. Linearized plasmid containing the binding site was incubated with crude K562 extract, free DNA and protein-DNA complexes were separated from unbound proteins by gel filtration. The left panel illustrates that specific DNA-protein complexes can be readily purified from crude nuclear extracts and imaged in the electron microscope. The higher magnification of the image (right panel) gives an idea of the level of resolution attainable with this technique. Contour mapping of all protein-DNA complexes photographed place the protein(s) at the expected AAAGTCC site (see arrow, Figure 6). These preliminary results indicate that ESI should prove to be a useful tool for the study of specific globin DNA-protein interactions.

CONCLUSIONS AND FURTHER DIRECTIONS

These preliminary results indicate that there may be intron A specific factors within K562 cell nuclear extracts. We are currently using techniques such as _in vitro_ transcription to ascertain if these sites have any functional significance.

The ability to purify specific DNA-protein complexes from crude extracts should prove to be useful for the analysis of trans-acting factors bound to globin DNA by ESI. Multiple erythroid-specific and generic factors binding to several globin promoter and enhancer regions has been demonstrated; therefore, globin transcription and switching may result from a combinatorial action of several factors (see Nickol and Felsenfeld, 1988; Choi and Engel, 1988 and Wall et al.,1988). We are currently utilizing ESI to study the interaction of proteins with specific globin promoter and enhancer sites as well as the possible combinatorial interaction of these sites brought about by protein induced conformational changes in DNA structure.

REFERENCES

Atweh G, Liu JM, Brickner HE, Zhu XX (1988). A silencer element from the alpha-globin gene inhibits expression of beta-like genes. Mol Cell Biol 8(11): 5047-5051.
Bazett-Jones DP (1988). Phosphorus imaging of the 7-S ribonucleoprotein particle. J Ultrastruct Molec Struct Res 99:59-69.
Bazett-Jones DP, Brown ML (1989). Electron microscopy reveals that the transcription factor TFIIIA bends 5S DNA. Mol

Cell Biol 1: in press

Bazett-Jones DP, Locklear L, Rattner JB (1988). Electron spectroscopic imaging of DNA. J Ultrastruct Molec Struct Res 99:48-58

Bazett-Jones DP, Ottensmeyer FP (1981). Phosphorus distribution in the nucleosome. Science 211:169-170.

Bazett-Jones DP, Yeckel M, Gottesfeld JM (1985) Nuclear extracts from globin-synthesizing cells enhance globin transcription in vitro. Nature 317(6040): 824-828.

Bornstein P, McKay J, Liska DJ, Apone S, Devarayalu S (1988). Interactions between the promoter and first intron are involved in transcriptional control of alpha1(I) collagen gene expression. Mol Cell Biol 8(11):4851-4857.

Choi OR, Engel JD (1988). Developmental regulation of beta-globin gene switching. Cell 55:17-26.

Galas DJ, Schmitz A (1978). DNase1 footprinting a simple method for the detection of protein-DNA binding specificity. Nucleic Acids Res 5(9):3157-3170.

Galson DL, Houseman DE (1988). Detection of two tissue-specific DNA-binding proteins with affinity for sites in the mouse beta-globin intervening sequence 2. MCB 8(1):381-392.

Gutierrez-Hartmann A, Siddiqui S, Loukin S (1987). Selective transcription and DNase1 protection of the rat prolactin gene by GH3 pituitary cell-free extracts. Proc Natl Acad Sci USA 84:5211-5215.

Hendrickson W (1985) Protein-DNA interactions studied by the gel electrophoresis-DNA binding assay. Biotechniques 3:198-207.

Killen PD, Burbelo PD, Martin GR, Yamada Y (1988) Characterization of the promoter for the alpha1(I)collagen gene. J Biol Chem 263(25):12310-12314.

Lanfranchi G, Linder S, Ringertz N (1984). Globin synthesis in heterokaryons formed between chick erythrocytes and human K562 cells or rat L6 myoblasts. J Cell Sci 66: 309-319

Miller AD, Bender MA, Harris EA, Kaleko M, Gelinas RE (1988). Design of retrovirus vectors for the transfer and expression of the human beta-globin gene. J Virol 62(11): 4337-4345.

Montovani R, Malgaretti N, Nicolis S, Ronchi A, Giglioni B, Ottolenghi S (1988). The effects of HPFH mutations in the human gamma-globin promoter on binding of ubiquitous and erythroid specifc nuclear factors. Nucleic Acids Res 16(16):7783-7797.

Muller M, Gerster T, Schaffner W (1988). Enhancer sequences

and the regulation of gene transcription. Eur J Biochem 176: 485-495.

Nickol JM, Felsenfeld G (1988). Biodirectional control of the chicken beta- and epsilon-globin genes by a shared enhancer. Proc Natl Acad Sci USA 85:2548-2552.

Superti-Furga G, Barberis A, Schaffner G, Busslinger M (1988). The -117 mutation in greek HPFH affects the binding of three nuclear factors to the CCAAT region of the gamma-globin gene. EMBO J 7(10):3099-3107.

Wada Y, Noguchi T (1988). In vitro differential expression of human globin genes. J Biol Chem 263(24): 12142-12146.

Wall L, deBoer E, Grosveld F (1988). The human beta-globin gene 3' enhancer contains multiple binding sites for an erythroid-specific protein. Genes and Development 2:1089-1100

Wingender E (1988). Compilation of transcription regulating proteins. Nucleic Acids Res 16(15):1879-1902.

Hemoglobin Switching, Part A: Transcriptional Regulation, pages 301–311
© 1989 Alan R. Liss, Inc.

IN VITRO TRANSCRIPTION WITH K562 CELL NUCLEAR EXTRACT AND GLOBIN GENES.

Yuko Wada and Constance Tom Noguchi

Laboratory of Chemical Biology, NIDDK, National Institutes of Health, Bethesda, Maryland 20892

SUMMARY

Cloned human epsilon-, A/gamma- and beta-globin genes, the insulin gene and the adenovirus 2 major late promoter (Ad2MLP) were employed for transcription *in vitro* with K562 nuclear extracts. Nuclear extracts could direct accurate initiation of transcription from epsilon-globin without supplement of a whole cell extract. A clear dependence of protein concentration of nuclear extracts on transcriptional enhancement was observed with the epsilon-globin gene. To examine the cis-acting DNA sequences 5' of the promoter region which may be important in specific expression of globin genes, the epsilon-globin template was truncated using restriction enzymes. Transcriptional activity of the epsilon-globin gene varied according to the truncation suggesting possible regions to which nuclear proteins involved in transcription may bind. Fractionation of nuclear extracts by ion exchange chromatography indicated that activity could be recovered in a 175 mM ammonium sulfate fraction while the 50 mM ammonium sulfate fraction decreased transcription activity. A/gamma globin gene and Ad2MLP could be transcribed in nuclear extracts at higher concentrations, however, beta-globin and the insulin gene were not transcribed at any concentration assayed, either from induced or uninduced cells. Transcription of the beta-globin gene was observed *in vitro* when K562 nuclear extracts were supplemented with HeLa whole cell extracts.

INTRODUCTION

Human alpha- and beta-like globin gene families have coordinate programs for each gene expression. Any abnormalities in the programs such as the timing of switching and the level of production of specific globins during different developmental stages cause a variety of hemoglobinopathies. The mechanism of hemoglobin switching is particularly relevant to beta-thalassemia and sickle cell anemia in which expression of the adult beta-globin polypeptide is abnormal. In these genotypes, the gamma-globin genes are usually intact and normal, and several therapeutic strategies have concentrated on reversing the developmental hemoglobin switch (from fetal to adult hemoglobin) to induce the production of normal gamma-globin polypeptide in place of abnormal or deficient beta-globin polypeptide.

The K562 human erythroleukemia cell line provides a model system to investigate the mechanism, of human hemoglobin switching. The cells express predominantly embryonic and fetal hemoglobins, but cannot express adult hemoglobin (Lozzio and Lozzio, 1975). K562 cells can be further induced for hemoglobin production by chemical stimuli such as hemin (Dean et al., 1985). Regulation of globin gene expression in K562 cells occurs at the level of transcription and much experimental evidence has pointed to the regulatory role of trans-acting factors and cis-acting DNA sequences in globin gene expression (Young et al, 1985; Fordis et al, 1986; Rutherford and Nienhuis, 1987).

The soluble cell-free *in vitro* transcription system provides us with a useful approach for studying the regulation of gene expression at the transcriptional level and to identify protein or other factors required for RNA production (Matsui et al., 1980). Whole cell extracts (Manley et al., 1980), cytoplasmic extracts supplemented with RNA polymerase II (Weil et al., 1979) and nuclear extracts (Dignam et al., 1983) have been used to simulate transcription as it occurs within the intact cell. HeLa cell extracts have been especially useful in demonstrating accurate initiation of transcription of cloned genes. Indeed, many globin genes have been shown to be correctly initiated in the HeLa whole cell extract system (Proudfoot et al., 1980; Spritz et al., 1981; Orkin et al., 1981). Supplementation of HeLa cell extract with induced (but not uninduced) K562 nuclear extract increases the efficiency for globin gene transcription (Bazett-Jones et al., 1985).

We have used the *in vitro* transcription system as a complete homologous system to examine the specificity of K562 extracts for globin gene transcription. K562 nuclear extracts and cloned human globin genes were used to investigate the cis-acting regulatory DNA regions 5' to the polypeptide coding region and to isolate and characterize the trans-acting regulatory proteins or factors involved in the transcriptional process. We prepared nuclear extracts from hemin-induced and uninduced K562 cells for use in a soluble cell-free *in vitro* system (as a run-off assay).

IN VITRO TRANSCRIPTION

The in vitro transcription system described by Dignam *et al.* (1983) was adapted for K562 cells (Wada and Noguchi, 1988) and is illustrated below.

In vitro transcription of K562 & template DNA

Preparation of nuclear extract

Prepare nuclei from K562 erythroleukemia cells
↓
Isolate nuclear proteins using 0.42 M NaCl buffer
↓
dialize extract into 0.1 M KCl buffer

Assay condition

Nuclear extract & DNA template
60 mM KCl, 12 mM $MgCl_2$ buffer, ATP, CTP, UTP, ^{32}P-GTP
↓
$30^\circ C$, 60 min
↓
Analyze ^{32}P-RNA by denaturing agarose gel
↓
Autoradiography

K562 cells were grown in RPMI 1640 medium containing 10% fetal calf serum, Hepes, penicillin and streptomycin in the presence of glutamine. The cells which were induced with 20 μM hemin were grown continuously in long term culture (2 months or longer) in the presence of hemin without supplement of glutamine. Nuclear extracts from induced and uninduced K562 cells were prepared from nuclei fractionated in buffer containing 0.42 M NaCl. The assay for *in vitro* transcription was carried out using up to 50% by volume of nuclear extract in a total reaction volume of 50 µl. The DNA template was cleaved downstream from the cap site by appropriate restriction enzymes (fig. 1). The reaction was incubated at 30°C for one hour. The produced run-off RNA transcript was glyoxalated and analyzed by electrophoresis in a 1.4% endosmotic agarose gel. The size of the ^{32}P-labeled run-off RNA transcript was determined by radio-autography. Nuclear extracts made from both uninduced and induced K562 cells were able to support transcriptional activity driven by Ad2MLP (fig. 1a). The transcriptional activity was observed most efficiently for a DNA concentration of 0.5 µg per µl of reaction.

A critical dependence of transcriptional activity on protein concentration of nuclear extracts was observed in transcription from the epsilon-globin gene (fig. 1b). Nuclear extracts from induced cells showed the maximum enhancement of transcription from epsilon-globin gene promoter at concentrations between 0.5 and 0.7 mg/ml (Wada and Noguchi, 1988). Extracts from uninduced cells required two times higher protein concentration to obtain the same level of transcription activity as induced extracts.

Optimum transcriptional activity of A/gamma-globin gene occurred at higher protein concentrations than that observed for the epsilon-globin gene and more comparable to that observed for Ad2MLP. This reflects the situation *in situ* where the epsilon-globin gene is more transcriptionally active than the gamma-globin gene. No beta-globin transcripts are observed within the intact K562 cell or in the *in vitro* transcription system with K562 nuclear extracts (fig. 1c).

Figure 1. *In vitro* transcription using K562 nuclear extracts. (a) Transcription driven by the adenovirus 2 major late promoter (Ad2MLP) DNA template produces an RNA transcript of 379 bases. The protein concentration was varied from 1.5 to 4 mg/ml for Ad2MLP. (b) Transcription of the epsilon-globin gene produces a 764 base RNA transcript. The protein concentration was varied from 0.2 to 4 mg/ml. (c) The beta-globin DNA template cut with Bam H1 would produce a 476 base RNA transcript. None is observed when the protein concentration varied from 0.2 to 4 mg/ml. Alternative restriction cut with Eco R1 also did not produce any RNA transcript. Data reproduced from Wada and Noguchi, 1988.

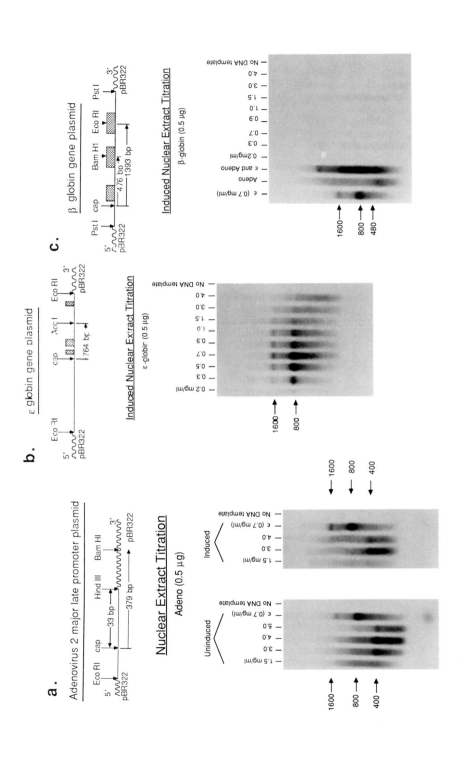

a.

Adenovirus 2 major late promoter plasmid

b.

ε globin gene plasmid

c.

β globin gene plasmid

TRANSCRIPTION FROM TRUNCATED EPSILON-GLOBIN DNA TEMPLATES

DNA templates for epsilon-globin of varying lengths up to 2,000 base pairs upstream from the cap site were compared for their activity in the *in vitro* transcription assay. The 5' truncated DNA templates were prepared by digesting 736 bp 3' downstream from the cap-site with Acc I and 5' upstream with a second restriction enzyme illustrated below.

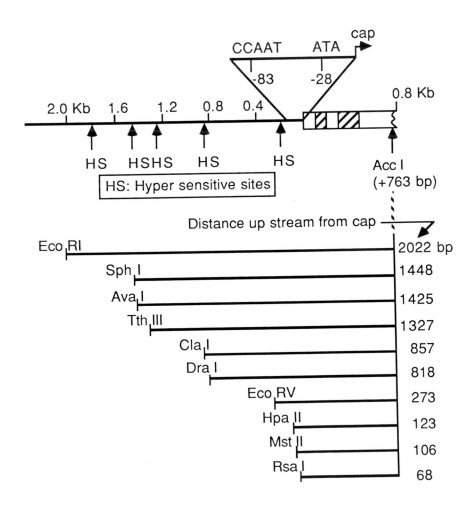

Preliminary results indicate that differential enhancement of transcription was obtained as the length of the promoter sequence varied. When the 3.7 kb Eco RI fragment of the epsilon-globin gene (including 2 kb 5' of the cap site) inserted into pBR322 (fig. 1) was cleaved as indicated above, the additional restriction fragments from pBR322 resulted in additional RNA transcripts of length other than the desired 736 base epsilon-globin transcript (Sassone-Corsi et al., 1981). These corresponded to end-end transcription of the pBR322 sequences and were observed in the agarose gel analysis of RNA transcripts. The same 3.7 kb fragment of the epsilon-globin gene inserted into puc19 gave rise to fewer restriction fragments and additional bands from end-end transcription was minimized (fig. 2).

To confirm that the truncated epsilon-globin DNA templates were correctly initiated, a ribonuclease T1 analysis of the run-off RNA transcripts was carried out. A ^{32}P-labeled RNA probe (fig. 3) was prepared by *in vitro* transcription using the SP6 promoter (Riboprobe from Promega Corporation). The unlabeled run-off RNA transcript was hybridized with the ^{32}P-labeled RNA probe and the hybridized material was digested with ribonuclease T1 (100 U/ml) and ribonuclease A (40 μg/ml). Ribonuclease resistant hybrids were analyzed by polyacrylamide electrophoresis and radioautography.

Accurate initiation of transcription was confirmed by the ribonuclease T1 analysis. The maximal transcriptional activity was observed using the DNA template containing 1451 bp upstream from the cap site (by Sph I digestion), consistent with results obtained when ^{32}P-labeled run-off transcripts were analyzed directly on agarose gels (from pBR322 and puc19 derived plasmids). In contrast, the activity dropped to a lower level when the DNA was digested with Ava I only 23 bp downstream from the Sph I site. Transcription was still detected when the DNA was cleaved by Rsa I, only 68 bp 5' from the cap-site, although accurate initiation with this template was not observed. Generally, no upstream initiation was detected and downstream initiation occurred when the template was 123 base pairs or less in length 5' from the cap-site.

TRANSCRIPTION OF BETA-GLOBIN WITH SUPPLEMENTATION BY HeLa EXTRACT

To determine whether the lack of expression of the beta-globin gene by K562 nuclear extracts was due to the absence or insufficient amount of transcription factors, or due to the presence of inhibitory factors specific for beta-globin gene expression we used a heterologous *in vitro* transcription system. A total of 1 μg of beta-globin gene DNA template in

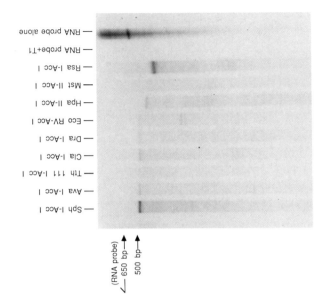

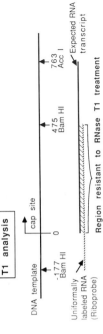

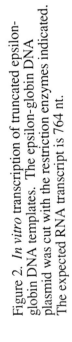

Figure 2. *In vitro* transcription of truncated epsilon-globin DNA templates. The epsilon-globin DNA plasmid was cut with the restriction enzymes indicated. The expected RNA transcript is 764 nt.

Figure 3. Ribonuclease T1 analysis of RNA transcripts produced from *in vitro* transcription of truncated epsilon-globin DNA templates. A uniformly labeled RNA probe (Riboprobe, Promega) corresponding to a Bam H1 fragment of the epsilon-globin gene was hybridized to the RNA transcript and digested with nuclease T1. A correctly initiated transcript gives rise to a 475nt protected RNA fragment.

a 50 µl reaction volume was transcribed by HeLa whole cell extract but not by K562 nuclear extract. However, when K562 nuclear extract was supplemented with HeLa whole cell extract, accurate transcription from the beta-globin gene was observed. The transcriptional activity of the supplemented K562 and HeLa system was lower in comparison with the HeLa whole cell extract alone. Hence, the K562 *in vitro* transcription system provides a method for the isolation of factors in HeLa cell extract which can activate beta-globin transcription.

FRACTIONATION OF NUCLEAR EXTRACT

For the identification and isolation of trans-acting factors for epsilon-globin gene expression, the K562 crude nuclear extract was fractionated by ion exchange chromatography using a stepwise elution with ammonium sulfate at 50 mM, 175 mM and 300 mM (No proteins were detected in the 1 M fraction). Only the 175 mM fraction (F175) was able to support transcription without the addition of more crude K562 nuclear extract. Furthermore, a lower protein concentration (0.2 mg/ml) of F175 yielded transcription of epsilon-globin comparable to that of 0.7 mg/ml to 1 mg/ml (optimum concentration) of crude nuclear extract. When combined with crude nuclear extract the 300 mM fraction (F300) had no effect on transcription activity while the 50 mM fraction (F50) decreased transcription activity. For F175, transcription activity decreased upon the addition of crude nuclear extract or F50. These results suggest the presence of inhibitory factors in crude extract or F50. Transcription activity of Ad2MLP was surprisingly low in F175 indicating specificity for globin gene trans-activation by this fraction.

DISCUSSION

Cloned human beta-like globin genes were accurately transcribed *in vitro* by extracts obtained from nuclei of K562 cells without the need of supplement with whole cell extracts or polymerase II. We demonstrated that the *in vitro* transcription system using K562 nuclear extracts was specific for epsilon- and gamma-globin genes, but the beta-globin gene promoter remained inactive, consistent with the lack of expression of beta-globin gene within the intact K562 cell. These results suggest that the *in vitro* transcription system can serve as an assay system to investigate the factors which contribute to the developmental activation or inactivation of globin genes. The specificity of the system for the epsilon-globin gene can be used to isolate proteins responsible for gene activation. Using reconstitution of protein fractions (such as F175) should allow us to determine which protein fractions are general transcription factors and which confers promoter specificity. Characterization of factors in the

HeLa cell extract which can activate beta-globin transcription may provide a source of factors which when used to supplement K562 nuclear extracts will activate other silent promoters. The *in vitro* transcription system provides a functional assay to complement the structural approach of studying proteins which bind to DNA directly.

REFERENCES

Bazett-Jones DP, Yeckel M, Gottesfeld JM (1985). Nuclear extracts from globin-synthesizing cells enhance globin transcription in vitro. Nature 317:824-828.

Dean A, Wu Y, Ley T, Fordis CM, Schechter AN (1985). Augmentation of hemoglobin synthesis by S-phase specific drugs in the K562 cell line. Prog Clin Biol Res 191:205-216.

Dignam JD, Lebovitz RM, Roeder RG (1983). Accurate transcription initiation by RNA polymerase II in a soluble extract from isolated mammalian nuclei. Nucleic Acids Res 11:1175-1189.

Fordis CM, Nelson N, McCormick M, Padmanabhan R, Howard B, Schechter AN (1986). The 5'-flanking sequences of human globin genes contribute to tissue specific expression. Biochem Biophys Res Commun 134:128-133.

Lozzio CB, Lozzio BB (1975). Human chronic myelogenous leukemia cell-line with positive Philadelphia chromosome. Blood 45:321-334.

Manley JL, Fire A, Cano A, Sharp PA, Gefter ML (1980). DNA-dependent transcription of adenovirus genes in a soluble whole-cell extract. Proc Natl Acad Sci USA 77:3855-3859.

Matsui T, Segall J, Weil PA, Roeder RG (1980). Multiple factors required for accurate initiation of transcription by purified RNA polymerase II. J Biol Chem 255:11992-11996.

Orkin SH, Goff SC (1981). Nonsense and frameshift mutations in beta[0]-thalassemia detected in cloned beta-globin genes. J Biol Chem 256:9782-9784.

Proudfoot NJ, Shander MHM, Manley JL, Fegter ML, Maniatis T (1980). Structure and in vitro transcription of human globin genes. Science 209:1329-1336.

Rutherford T, Nienhuis AW (1987). Human globin gene promoter sequences are sufficient for specific expression of a hybrid gene transfected into tissue culture cells. Mol Cell Biol 7:398-402.

Sassone-Corsi P, Corden J, Kedinger C, Chambon P (1981). Promotion of specific in vitro transcription by excised "TATA: box sequences inserted in a foreign nucleotide environment. Nucleic Acids Res 9:3941-3958.

Spritz RA, Jagadeeswaran P, Choudary PV, Biro PA, Elder JT, de Riel JK, Manley JL, Gefter ML, Forget BG, Weissman SM (1981). Base substitution in an intervening sequence of a beta[+]-thalassemic human globin gene. Proc Natl Acad Sci USA 78:2455-2459.

Wada Y, Noguchi CT (1988). *In vitro* differential expression of human globin genes. J Biol Chem 263:12142-12146.

Weil PA, Luse DS, Segall J, Roeder RG (1979). Selective and accurate initiation of transcription at the Ad2 major late promoter in a soluble system dependent on purified RNA polymerase II and DNA. Cell 18:469-484.

Young K, Donovan-Peluso M, Cubbon R, Bank A (1985). Trans acting regulation of beta globin gene expression in erythroleukemia (K562) cells. Nucleic Acids Res 13:5203-5213.

Hemoglobin Switching, Part A: Transcriptional Regulation, pages 313–321

CLONING OF cDNA FROM INDUCED K562 CELLS WHICH CAN ACTIVATE GLOBIN GENE EXPRESSION.

Yongji Wu and Constance Tom Noguchi

Laboratory of Chemical Biology, NIDDK, National Institutes of Health, Bethesda, Maryland 20892

SUMMARY

A cDNA library from induced K562 cells was constructed and differentially screened for the isolation of clones encoding trans-acting factors which increase globin gene expression. The current study assumes that induced K562 cells contain transcriptionally active factors specific for globin genes which are absent or present only at very low levels in uninduced K562 cells. Upon screening the recombinant library, 75 cDNA clones hybridized specifically with cDNA probes from induced K562 cells and hybridized only slightly or not at all with cDNA probes from uninduced K562 cells and HL-60 cells, or from globin genes. Forty-five of the cDNA clones were full length complements to the corresponding RNA. To screen for trans-acting factors which can activate globin gene expression, the cDNA clones were inserted into a eukaryotic expression vector and co-transfected into HeLa cells with another vector containing the epsilon globin promoter 5' to the bacterial CAT (chloramphenicol acetyl transferase) reporter gene. Partial screening of the 45 full length clones revealed one cDNA (clone #17) that was able to increase CAT activity (driven by the epsilon-globin promoter) by about 2.5 times. This cDNA is 522 nucleotides in length and contains a long open reading frame. Examination with known DNA sequences indicates >95% homology with the ferritin heavy chain.

INTRODUCTION

The specific transcription of human globin genes may involve the complex interaction of a variety of factors (Karlson and Nienhuis, 1985). For the study of globin gene expression, we use the K562 human erythroleukemia cell line as a model system for globin gene expression. The K562 cell line can be induced by hemin to accumulate embryonic and fetal hemoglobin, but not adult hemoglobin (Rutherford et al, 1979). It has been demonstrated that β-globin gene is intact but inactive in these cells (Fordis et all, 1984). The ζ-globin gene promoter functions after microinjection into oocytes but not after transfection into HeLa or COS cells (Proudfoot et al., 1984), suggesting that there may be transcriptional factors specific for embryonic globin genes. The goal of the present study is to clone and characterize such factors.

Following hemin induction, human embryonic hemoglobins (Gower I ($\zeta_2\epsilon_2$), Gower 2 ($\alpha_2\epsilon_2$) and Portland ($\zeta_2\gamma_2$)) and fetal hemoglobin ($\alpha_2\gamma_2$), but not adult hemoglobin ($\alpha_2\beta_2$) are accumulated in induced K562 cells. If there are any trans-acting factors involved in the induction of ε-, ζ-, γ- or α-globin genes, these protein factors must be present in induced K562 cells, but absent or only at low level in uninduced K562 cells. For the purpose of search for these protein factors, we have constructed a cDNA library from mRNA of long-term (2 years) induced K562 cells and have screened 1.5×10^5 recombinants with ^{32}P-labeled cDNA probes from induced and uninduced K562 cells. Those cDNA clones which are differentially expressed have been further screened with ^{32}P-labeled embryonic and fetal globin gene DNA or cDNA probes as well as ^{32}P-labeled cDNA probes from HL-60 (Collins et al., 1977) (human promyelocytic leukemia cell line) cells to subtract cDNA clones corresponding to embryonic and fetal globin as well as proteins present in both the induced K562 cells and HL-60 cells. To determine whether those differentially expressed cDNA clones were full length complements to the corresponding RNA, a Northern blot analysis was carried out (Thomas, 1980). The cDNA clones in full length were inserted into an expression vector driven by the SV40 promoter and their functional effect on the activation of ε-globin gene promoter was examined by co-transfection into HeLa cells with another vector containing the bacterial CAT gene driven by the ε-globin gene promoter. cDNA clones which increased expression of the CAT gene in HeLa cells were sequenced and further analyzed.

CONSTRUCTION OF THE cDNA LIBRARY.

The cytoplasmic RNAs were prepared from K562 cells induced by 20 μM hemin for long term (2 years) (Berger and Kimmel, 1987). Poly(A)$^+$ mRNAs were obtained from the cytoplasmic RNAs by chromatography on oligo (dT)-cellulose columns, and the mRNA was copied into double-stranded cDNA using reverse transcriptase, ribonuclease H, DNA polymerase and T4 DNA polymerase (cDNA synthesis system,

Amersham). This procedure was also used to prepare cDNA probes from uninduced and induced K562 cells and HL60 cells. About 0.5 μg of double-stranded cDNA was synthesized and fractionated by chromatography on a Sephadex 4B column. cDNAs larger than 400 base pairs were isolated and methylated to protect internal Eco R1 sites and ligated with Eco R1 linkers. After digesting with Eco R1 to generate cohesive ends and separation from excess linkers by chromatography, the cDNAs were ligated to λgt10 arms and packaged *in vitro*. The packaged recombinant lambda phage was infected into host strain (NM$_{514}$) to replicate enough copies of recombinant phage for differential screening. The cDNA library from induced K562 cells consisted of 4.5 x 10^5 independent recombinants.

DIFFERENTIAL SCREENING OF THE cDNA LIBARY.

About 1.5 x 10^5 recombinant clones were differentially screened (Figure 1). ^{32}P-labeled cDNA probes from mRNAs of uninduced, induced K562 cells and HL-60 cells, and ^{32}P-labeled human ε , ζ–, δ–, α– and γ-globin genomic or cDNA probes (prepared by random hexamer primer extension (Feinberg and Vogelstein, 1983) to a specific activity of 1-3x10^9 CPM/μg) were used to screen cDNA library as illustrated below. Duplicate nitrocellulose filters were prehybridized at 42°C in 50% formamide, 5X Denhardt's solution, 5xSSPE, 0.1% SDS, 100 μg/ml denatured salmon sperm DNA and 1 μg/ml poly (A). Hybridization was performed in a similar solution containing cDNA probe at a specific activity of 1x10^6 CPM/ml at 42°C overnight. Filters were washed twice with 1xSSC and 0.1% SDS at 42°C and once with 0.2xSSC and 0.1%SDS at 65°C. Most of the cDNA clones were found to be non-specific of induced K562 cells. Seventy-five cDNA clones were found to be specific for induced K562 cDNA. That is, they hybridized weakly or not at all with cDNA from uninduced K562 or HL-60 cells, or with the globin probes (Figure 2). Several clones hybridized more strongly with cDNA from uninduced K562 cells. These most likely represented genes which were down-regulated during induction.

NORTHERN BLOT ANALYSIS FOR FULL LENGTH cDNA.

The 75 cDNA clones found to be specific for induced K562 cells were examined to determine if they represented full length cDNA. RNA was prepared from uninduced and induced K562 cells. For northern blot analysis, 20μg of each RNA was separated by electrophoresis through glyoxal-DMSO/agarose gel and transferred onto nitrocellulose filters. The

DIFFERENTIAL SCREENING

From cDNA library (4.5×10^5 recombinants), 1.5×10^5 clones screened.

1. Primary Screening:

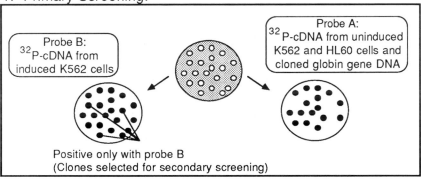

Probe B:
^{32}P-cDNA from induced K562 cells

Probe A:
^{32}P-cDNA from uninduced K562 and HL60 cells and cloned globin gene DNA

Positive only with probe B
(Clones selected for secondary screening)

2. Secondary Screening:

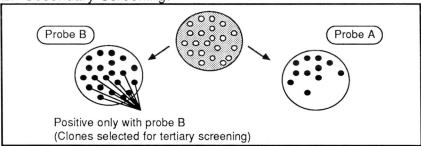

Probe B

Probe A

Positive only with probe B
(Clones selected for tertiary screening)

3. Tertiary Screening:

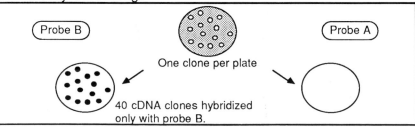

Probe B

Probe A

One clone per plate

40 cDNA clones hybridized only with probe B.

Figure 1. Schematic for differential screening of cDNA clones.

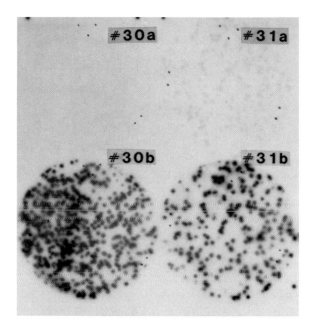

Figure 2. Autoradiogram of tertiary screening of cDNA specific for induced K562 cells. 30a) Clone 30 does not hybridize with probe A containing cDNA from uninduced K562 and HL60 cells and globin DNA. 30b) Clone 30 hybridizes strongly with probe B containing cDNA from induced K562 cells. 31a) Clone 31 hybridizes weakly with probe A. 31b) Clone 31 hybridizes strongly with probe B.

blot was baked for 2 hr at 80°C in vacuum. The blotted filters were prehybridized in 50% formamide, 5xSSC, 50 mM sodium phosphate buffer, pH 6.5, 250 µg/ml denatured salmon sperm DNA and 10X Denhardt's solution at 42°C overnight. Hybridization was performed in a similar solution but containing ^{32}P-labeled cloned cDNA at 42°C for 20hr. Filters were washed four times with 2xSSC AND 0.1% SDS at 50°C. Forty-five cDNA clones were found not only to be specific for induced K562 cells but also to be full length complements to the corresponding RNAs.

ASSAY FOR TRANS-ACTIVATION OF ε-GLOBIN GENE.

Cloned cDNAs were separated from λgt10 by digestion with Eco RI and inserted into a eukaryotic expression vector derived from the Okayama-Berg cloning vectors (Okayama and Berg, 1983) which contains the SV40 promoter and other segments necessary for the expression of inserted cDNA in mammalian cells and for replication and selection of the plasmid in E. coli. For the reporter gene system, ε-globin gene promoter region consisting of 1.46 kb of the 5' flanking region upstream and including the cap site was linked to the bacterial CAT gene (pεCAT).

To assay the trans-acting transcriptional activity of the cDNA clone, the cDNA expression plasmid and the globin-CAT plasmid were co-transfected into HeLa cells. Approximately 5×10^5 cells were plated on 10cm tissue culture dish and allowed to grow for 24 hr. The culture medium was then replaced with fresh medium (DMEM). 10 μg of pεCAT and 10μg of cDNA plasmid were added to the cells 3 hr later. Cells were incubated for 4 hr, washed in serum-free medium twice, fed with complete medium and incubated for 48 hr. Lysates were prepared from harvested cells and assayed for CAT activity (Gorman, et al., 1982). After partial screening of the cDNA clones, most of the clones assayed exhibited only a background level of CAT activity (Figure 3).

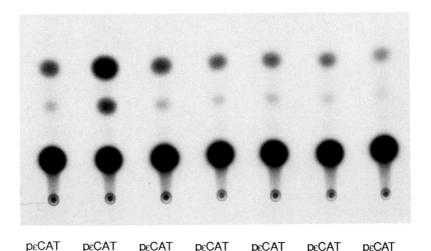

pεCAT	pεCAT	pεCAT	pεCAT	pεCAT	pεCAT	pεCAT
+ #10	+ #17	+ #13	+ #14	+ #16	+ #19	+ #28

Figure 3. CAT assays of cotransfection of cDNA clones and pεCAT into HeLa cells. Only clone #17 shows an increase

(three fold) above control values.

Clone #17 increased CAT activity about 2.5 fold (Figure 3) indicating an increase in CAT production driven by the ε-globin promoter. To determine the nature of the polypeptide coded for by the cDNA, clone #17 was subcloned into puc18 and the double stranded plasmid was sequenced by the dideoxy chain - termination method of Sanger et al. (1977). Clone #17 was 522 base pairs and shared a high degree of homology (499 matches out of 502 nucleotides (99.4%)) with the ferritin heavy chain cDNA (Figure 4).

CONCLUSION

Hemoglobin switching during development reflects the sequential expression of the individual genes within the globin gene clusters. The study of globin gene transcription may contribute to the elucidation of the developmental control of gene expression. Understanding of the regulatory mechanism of human globin gene expression should provide insight into some hematological disorders such as thalassemia. Further manipulation of globin gene expression may offer new therapeutic approaches to patients with severe β-thalassemia or sickle cell anemia, or other types of hemoglobinopathies.

The current study has provided us with several cDNA clones specific for induced K562 cells. Several of these, in addition to #17, have been identified for their ability to increase ε-globin gene promoter activity. These clones will be further investigated for their effect on other globin promoters and for the precise mechanism by which they affect globin gene transcription.

```
          10        20        30        40        50        60        70        80
TGAAGCTGCAGAACCAACGAGGTGGCCGAATCTTCCTTCAGGATATCAAGAAACCAGACTGTGATGACTGGGAGAGCGCG
TGAAGCTGCAGAACCAACGAGGTGGCCGAATCTTCCTTCAGGATATCAAGAAACCAGACTGTGATGACTGGGAGAGCGGG
         300       310       320       330       340       350       360

          90       100       110       120       130       140       150       160
GTGAATGCAATGGAGTGTGCATTACATTTGGAAAAAAATGTGAATCAGTCACTACTGAACTGCACAAACTGGCCACTGA
CTGAATGCAATGGAGTGTGCATTACATTTGGAAAAAAATGTGAATCAGTCACTACTGAACTGCACAAACTGGCCACTGA
         380       390       400       410       420       430       440

         170       180       190       200       210       220       230       240
CAAAAATGACCCCCATTGTGTGACTTCATTGAGACACATTACCTGAATGAGCAGGTGAAAGCCATCAAAGAATTGGGTG
CAAAAATGACCCCCATTGTGTGACTTCATTGAGACACATTACCTGAATGAGCAGGTGAAAGCCATCAAAGAATTGGGTG
         460       470       480       490       500       510       520

         250       260       270       280       290       300       310
ACCACGTGACCAACTTGCGCAAGATGGGAGCGCC-GAATCTGGCTTGGCGGAATATCTCTTTGACAAGCACACCCTGGGA
ACCACGTGACCAACTTGCGCAAGATGGGAGCGCCCGAATCTGGCTTGGCGGAATATCTCTTTGACAAGCACACCCTGGGA
         540       550       560       570       580       590       600

         330       340       350       360       370       380       390
GACAGTGATAATGAAAGCTAAGCCTCGGCTAATTTCCCATAGCCGTGGGGTGACTTCCCTGGTCACCAAGGCAGTGCA
GACAGTGATAATGAAAGCTAAGCCTCGGCTAATTTCCCATAGCCGTGGGGTGACTTCCCTGGTCACCAAGGCAGTGCA
         620       630       640       650       660       670       680

         410       420       430       440       450       460       470
TGCATGTTGGGGTTTCCTTTACCTTTCTATAAGTTGTACCAAAACATCCACTTAAGTTCTTTGATTTGTACCATTCCTT
TGCATGTTGGGGTTTCCTTTACCTTTCTATAAGTTGTACCAAAACATCCACTTAAGTTCTTTGATTTGTACCATTCCTT
         700       710       720       730       740       750       760

         490       500
CAAATAAAGAAATTTGGTACCC
CAAATAAAGAAATTTGGTACCC
         780       790
```

Figure 4. Sequence homology between clone #17 (upper) and human ferritin heavy chain mRNA (Boyd et al., 1985) (lower). Mismatches are indicated as bold face letters. The homology is 99.4% (or 499/502 nucleotides).

REFERENCES

Berger SL, Kimmel AR (1987). "Guide to molecular cloning techniques." San Diego: Academic Press, pp 231-232.

Boyd D, Vecoli C, Belcher DM, Jain SK, Drysdale JW (1985). Structural and functional relationships of human ferritin H and L chains deduced from cDNA clones. J Biol Chem 260:11755-11761.

Collins SJ, Gallo RC, Gallogher RE (1977). Continuous growth and differentiation of human myeloid leukemic cells in suspension culture. Nature 270: 347-349.

Feinberg AP, Vogelstein B (1983). A technique for radiolabeling DNA restriction endonuclease fragments to high specific activity. Ana Biochem 132:6-13.

Fordis MC, Anagnou NP, Dean A, Nienhuis AW, Schechter AN (1984). A β-globin gene, inactive in the K562 leukemic cells, functions normally in a heterologous expression system. Proc Natl Acad Sci USA 81:4485-4489.

Gorman CM, Moffat LF, Howard BH (1982). Recombinant genomes which express chloramphenical acetyl transferase in mammalian cells. Mol Cell Biol 2:1044 1051.

Karlson S, Nienhuis AW (1985). Developmental regulation of human globin genes. Ann Rev Biochem 54:1031-1082.

Okayama H, Berg P (1983). A cDNA cloning vector that permits expression of cDNA inserts in mammalian cells. Mol Cell Biol 2:280-289.

Proudfoot NJ, Rutherford TR, Partington GA (1984). Transcriptional analysis of human zeta globin genes. EMBO J 3:1533-1540.

Rutherford TR, Clegg JB, Weatherall DJ (1979). K562 human leukemic cells synthesise embryonic haemoglobin in response to haemin. Nature 280:164165.

Sanger F, Nicklen S, Coulson AR (1977). DNA sequencing with chain-terminating inhibitors. Proc Natl Acad Sci USA 74:5463-5467

Thomas PS (1980). Hybridization of denatured RNA and small DNA fragment transferred to nitrocellulose. Proc Natl Acad Sci USA 77:5201-5205.

Hemoglobin Switching, Part A: Transcriptional Regulation, pages 323–333

THE GLOBIN SWITCH AT THE LEVEL OF mRNA IN THE DEVELOPING MOUSE

E. Whitelaw, P. Lamb, P. Hogben and N.J. Proudfoot

Sir William Dunn School of Pathology, University of Oxford, Oxford OX1 3RE, UK

ABSTRACT

We have carried out a detailed analysis of the relative amounts of ζ, α, βH1, εy^2 and adult β globin mRNA in different tissues of the mouse embryo from Day 8.5 to Day 17.5 i.e. from the first signs of erythropoiesis until almost the end of gestation (birth is Day 19). Interestingly, we find that the ζ to α "switch" occurs 24 hours earlier in yolk sac than it does in fetal liver and that the ratio of ζ to α mRNA remains higher in the peripheral blood than in the yolk sac or the fetal liver during the latter half of gestation. In fact, ζ mRNA remains present in peripheral blood until Day 15.5. The switch at the mRNA level appears to mimic that found by others [Popp et al, 1987] at the protein level, at least for peripheral blood. This suggests that regulation is not occurring to any major extent at the translational level. The reiteration of the switch in fetal liver suggests that local environmental factors are involved.

We find two switches within the β cluster; βH1 is expressed first, then εy^2 and lastly the two adult β globins (β major and β minor). The switch from βH1 to εy^2 occurs around Day 11.5 and the switch from εy^2 to the adult β globins occurs around Day 15.5. This means that the pattern of expression of the β-like globin genes in the mouse is more analogous to that found in the human than was previously thought.

INTRODUCTION

The ontogenic development of erythropoiesis in mammals is characterized by sequential changes in haemoglobin synthesis, erythropoietic site and erythroblast morphology [for review, see Wood, 1982]. These changes (or "switches") provide us with an interesting model with which to investigate the mechanisms underlying the switches in gene expression at the molecular level. The most complete studies of the changes in haemoglobin synthesis during development have been carried out in the human [for review, see Wood, 1982]. These studies all describe the "switch" from embryonic to fetal and from fetal to adult Hb at the protein level [Wood, 1976; Peschle et al, 1985]. A detailed study at the level of mRNA has not been carried out. It seemed important to us to determine whether or not these switches in expression were indeed occurring at the transcriptional level rather than at the translational level. In order to cover all stages of gestation, where the date of conception was well defined, we decided to carry out this study in the mouse.

The globin genes of the mouse, like those of man, lie in two clusters; the β like cluster on chromosome 7 [Jahn et al, 1980; Leder et al, 1980] and the α-like cluster on chromosome 11 [Leder et al, 1985], see Figure 1. The genes coding for the β-like embryonic globins (εy^2, βH0, βH1) lie in a cluster 5' to the two adult β genes, β^{maj} and β^{min}. The embryonic α-like globin, ζ, lies 5' to the two adult α genes. The globin gene nomenclature for the mouse is complex. The major problem is that, for historical reasons, the names of the protein products do not coincide with the name of the genes which produce them. There are three embryonic haemoglobins: EI (x_2y_2), EII (α_2y_2) and EIII (α_2z_2) and two adult haemoglobins ($\alpha_2\beta_2^{maj}$ and $\alpha_2\beta_2^{min}$). The relative amount of β^{maj} to β^{min} varies from strain to strain and at different times of development [Whitney, 1977]. The gene, εy^2, produces the globin called y. The gene, βH1, produces the globin called z, and the gene, βH0, is thought to be a functional gene but the levels of globin mRNA produced are extremely low [Hill et al, 1984]. The embryonic α-like globin, ζ, sometimes called x, produces the globin called x.

The mouse, like most vertebrates, has four primary sites of erythropoiesis: yolk sac, liver, spleen and bone marrow. Firstly, between Days 8 and 14 of gestation, the yolk-sac blood islands give rise to the primitive nucleated erythrocytes. These primitive nucleated erythrocytes are replaced by a population of enucleated erythrocytes that differentiate in the second site of erythropoiesis, the fetal liver. These liver-derived erythrocytes first appear at Day 10. The liver continues to function as an erythropoietic tissue until 10 days after birth. We have analysed the relative amounts of ζ, α, βH1, εy^2 and β major and minor globin mRNA in different erythroid tissues of the mouse embryo (yolk sac, fetal liver, peripheral blood) from the first signs of erythropoiesis (Day 8.5) until almost the end of gestation (Day 17.5). Birth in the mouse strain we used is at Day 19.

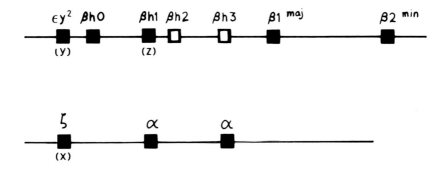

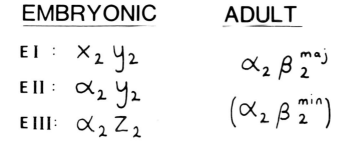

Figure 1. Mouse globin gene organization. A line diagram representing the β-cluster on chromosome 7 [Jahn et al, 1980] and the α-cluster on chromosome 11 [Leder et al, 1985]. The composition of the embryonic and adult haemoglobin tetramers are shown below.

METHODS

Mouse Matings and Dissection

All mouse embryos used in this study were derived from matings between outbred P.O. (Pathology, Oxford) parents. Matings were assumed to occur at the mid-point of the dark cycle (midnight). The morning of finding the plug was designated 0.5 day gestation. Pregnant females were killed by cervical dislocation between 10-12 a.m. on Days 8.5-17.5. Embryos were dissected free of decidua and the placenta was removed. Peripheral blood was collected after decapitation. Yolk sac was collected from Day 9.5 onwards. Fetal livers were dissected from the embryo from Day 10.5 onwards since at Day 9.5 no hepatic anlage was present.

RNA preparation and analysis

RNA from embryonic tissues was prepared using the guanidinium isothiocyanate procedure [Glisin et al, 1974]. The levels of α-globin, ζ-globin, βH1-globin and adult β-globin mRNA were determined using S1 nuclease. The probe used to determine mouse α-globin mRNA was double-stranded end-labelled at the BstEII site in exon 3. The DNA used was obtained by linearising a plasmid containing the mouse α-globin gene [Nishioka & Leder, 1979] with BstEII. The probe used to determine the mouse ζ-globin mRNA was double-stranded end-labelled at the AvaII site in exon 3 of the ζ-globin gene [Leder et al, 1985]. The probe used to determine the mouse β-globin gene mRNA was a Pst-NcoI fragment double-stranded end-labelled from exon 2 [Konkel et al, 1979]. This probe does not differentiate between β major and β minor mRNA. Therefore, the levels detected represent the sum of both. The probe used to determine the levels of βH1 mRNA was double-stranded and end-labelled at the NcoI site in exon 2 [Hill et al, 1984].

The level of ϵy^2 mRNA [Hansen et al, 1982] was determined using uniformly labelled RNA probes synthesized with SP6 polymerase according to Melton et al [1984]. The βH1 cDNA clone was kindly provided by Andy McMahon (Roche Institute) [Wilkinson et al, 1988]. pSP65Mε [Baron & Maniatis, 1986] was linearised with HindIII. It contained the first exon of the ϵy^2 gene [Hansen et al, 1982].

In situ hybridization

Procedures for *in situ* hybridization were as described by Cox et al [1984] with modifications as described by Wilkinson et al [1988]. [^{35}S]UTP-labelled single-stranded sense and antisense RNA probes were prepared by standard procedures [Melton et al, 1984]. Autoradiography was performed using Ilford K5 nuclear track emulsion. Finally, sections were stained with haematoxolin and eosin.

RESULTS and DISCUSSION

Alpha globin cluster

In Figure 2, we have represented the levels of ζ and α globin mRNA in yolk sac, blood and fetal liver at various days of gestation as a ratio of ζ to α. The ratio of ζ to α mRNA could be calculated directly knowing the specific activity of the two probes. A number of interesting things emerge. Firstly, the ζ to α switch at the mRNA level in the peripheral blood of the mouse is surprisingly gradual. Even at the earliest time point (9.5 days), the level of α globin mRNA is twice that of ζ. Samples of 8.5 day whole embryos (with decidua and placenta removed) have a ζ to α mRNA ratio which is

GLOBIN mRNA LEVELS

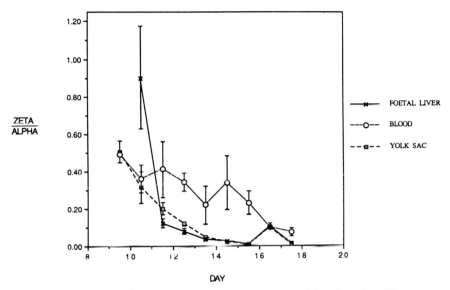

Figure 2. Ratio of ζ to α mRNA in fetal liver, blood and yolk sac from Day 9.5 to Day 17.5. Bars represent S.E.M. The probes used to measure the mRNA levels are described in the Methods section.

close to 1 (data not shown) but α globin mRNA appears to be present from the very start of erythropoiesis. Furthermore, ζ globin mRNA has not disappeared until Day 17.5. These data are comparable to those carried out previously at the protein level in the mouse [Popp et al, 1987]. However this is a much more gradual switch than that found in the human [Peschle et al, 1985].

Our results on the ζ to α mRNA ratio in the yolk sac show a clear switch during gestation (see Figure 2). This is consistent with the fact that primitive nucleated erythrocytes purified from the yolk sac have been shown to undergo a switch at the protein level in the mouse [Brotherton et al, 1979], the hamster [Boussios et al, 1982] and in the human [Peschle et al, 1985].

The ζ to α globin mRNA in the yolk sac drops more rapidly than in peripheral blood. Why the ratio in the yolk sac should be lower than that in the blood from Days 12.5-15.5 is not clear. Since circulating cells will include those which were produced a number of days earlier, then on any particular day we might expect to find a higher ζ to α ratio in peripheral blood than at the site of production, if the switching occurs less rapidly in blood than in the yolk sac.

Perhaps the most interesting finding is that the ζ to α "switch" occurs 24 hours earlier in yolk sac (Day 9.5) than it does in fetal liver (Day 10.5). The ζ to α ratio in fetal liver at Day 10.5, when the fetal liver is still a rudimentary organ, is very high (approaching 1), and the ratio in peripheral blood at the same moment in time is much lower (0.4). These results also suggest that there are local environmental effects. These may be acting at the transcriptional level. For example the levels of hypothetical growth factors in this tissue where cells are under-going rapid cell division may influence the rate of transcription of the globin genes. Alternatively, local environmental effects may be selecting for a particular sub-population of nucleated erythroid cells from the peripheral blood which have a higher ζ to α ratio. It should be noted that the nature of the induction of erythropoiesis in fetal liver remains controversial. Although most evidence suggests that this is a seeding event (resulting from the migration and seeding of yolk-sac divided cells), *de novo* induction from cells within the developing hepatic anlage remains a possibility. Our results do not directly address this problem. We do know from histology carried out on sectioned embryos at 10.5 days that many nucleated erythroid cells are present in the hepatic anlage (data not shown) and it should be noted that no attempt has been made to separate nucleated from enucleated erythroid cells.

Our data may suggest that when a tissue first becomes erythropoietic and is undergoing rapid cell division, the ζ to α ratio is at its highest. This is reminiscent of stress erythropoiesis in adult humans where it is well documented that under acute erythropoietic stress increased levels of globin from an earlier developmental stage (fetal globins) are produced [Stamatoyanno-poulos et al, 1987]. We are currently investigating whether the same phenomenon is true when erythropoiesis moves to its third site, the spleen (around Day 14.5).

Beta Globin Cluster

We have also analysed the same series of mRNA samples from peripheral blood with probes for the β-globin cluster: the embryonic genes, $\beta H1$ and ϵy^2 and the adult genes, β major and β minor. The results are represented in Figure 3 in the form of three overlying graphs where the level of mRNA of any particular gene is represented as a percentage of the total level of β-globin present at that time point. This can be calculated directly knowing the specific activity of all three probes.

We find that at Day 9.5, $\beta H1$ mRNA is by far the most predominant but by Day 11.5 its levels are lower than that of ϵy^2. The amount of ϵy^2 mRNA remains relatively high from Day 11.5 to Day 15.5. After Day 15.5 the majority of the β-like globin mRNA present in peripheral blood is of the adult type, i.e. β major and β minor.

MOUSE BETA GLOBIN CLUSTER

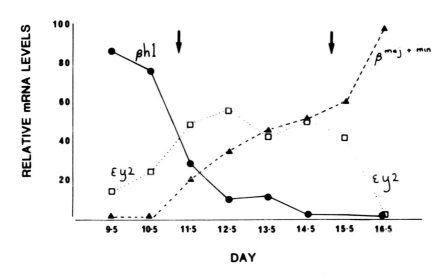

Figure 3.   Relative levels of mRNA of the genes from the β-cluster: βH1, εy² and adult β-globin (sum of β^{maj} and β^{min}). The analysis has been carried out on mRNA from peripheral blood from Day 9.5 to Day 16.5. The probes used are described in the Methods section.

This shows that the β-cluster genes of the mouse undergo two switches: the first from βH1 to εy² at Day 11.5 and the second from εy² to β major at Day 15.5. This is reminiscent of the switching in the human β-cluster, although the second switch in the mouse does appear to occur earlier, i.e. in the middle of the last trimester, rather than at the end of the last trimester. Some studies have been carried out previously comparing the levels of βH1 and εy² mRNA around Days 10.5-12.5, and these also reveal a switch in their relative amounts over this period [Farace et al, 1984; Chada et al, 1986].

In Situ Hybridization

In situ hybridization to RNA in sectioned mouse embryos has been used to detect the expression of various globin genes in situations where dissection is impossible due to the small number of cells concerned. Our preliminary experiments on Day 10.5 embryos reveal εy² mRNA in the mesodermal layer of yolk sac (see Fig.4A and 4B) and in the yolk sac derived nucleated erythroid cells

A.

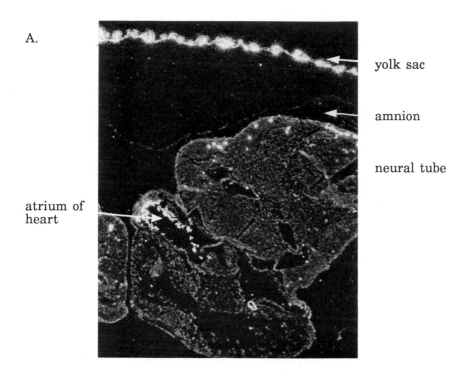

yolk sac

amnion

neural tube

atrium of
heart

B.

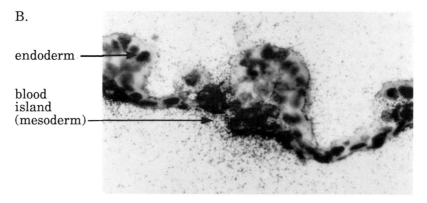

endoderm ——

blood
island
(mesoderm)——

Figure 4. *In situ* hybridization to a 10.5 Day mouse embryo. The probe used, εy², was uniformly labelled with ³⁵S-UTP (see Materials & Methods). A. Photograph of whole section under dark field. B. Enlarged photograph of yolk sac under light field.

resting in the chamber of the heart (see Fig. 4A and 5). The yolk sac is made up of two layers. The outside layer is derived from endoderm. Erythroid cells do not differentiate within this layer but these cells are thought to be required for erythropoiesis to occur in the adjacent layer (mesoderm) [Miura and Wilt, 1969]. The inner, mesodermal layer differentiates into either blood islands (erythropoietic) or connective tissue. Hybridization to the antisense ϵy^2 probe can be seen in localized areas within this mesodermal layer (Fig.4B).

One of the advantages of this technique is that we can detect mRNA at the level of single cells (see Figure 5). We also detect hybridization to small clusters of cells within the embryo proper. Visualization at higher magnification reveals that these represent nucleated erythroid cells lying within the developing blood vessels (angiogenesis starts around 9.5 days). Hybridization using the sense probe (which would hybridize to antisense transcripts) reveals no

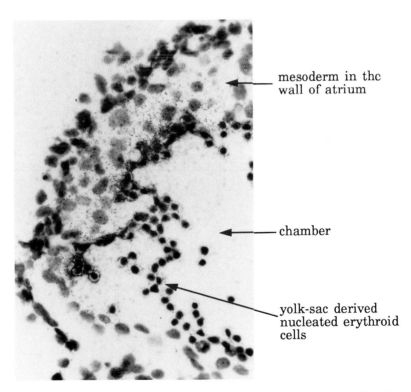

mesoderm in the wall of atrium

chamber

yolk-sac derived nucleated erythroid cells

Figure 5. *In situ* hybridization to atrium of heart at Day 10.5. The probe used was ϵy^2 uniformly labelled with ^{35}S-UTP (see Materials and Methods). Light field.

hybridization above background (data not shown) in any tissue. It is interesting that the positive cells within the blood islands appear to have a much higher level of labelling than that of the circulating nucleated erythrocytes. This may reflect their younger age, since we would expect the level of mRNA in the older, circulating cells to be dropping.

In situ hybridization in combination with the more quantitative analysis of RNA in solution will give us a good overview of erythropoiesis in the developing mouse. We are currently investigating the levels of various haemopoeitic growth factors during this period in an attempt to understand their role in the switches in globin gene expression.

ACKNOWLEDGEMENTS

We thank Andy McMahon (Roche Institute, New Jersey) for cDNA clones and for help with in situ hybridization. This work was supported by an M.R.C. project grant (no. G8708393CB).

REFERENCES

Baron M H and Maniatis T (1986). Rapid reprogramming of globin gene expression in transient heterokaryons. Cell 46, 591-602
Boussios T, Bertles J F and Clegg J B (1982). Simultaneous expression of globin genes for embryonic and adult hemoglobins during mammalian ontogeny. Science 218, 1225-1227
Brotherton T W, Chui D H K, Gauldie J and Patterson M (1979). Hemoglobin ontogeny during normal mouse fetal development. Proc.Natl.Acad.Sci.USA 76, 2853-2857
Chada K, Magram J and Costantini F (1986). An embryonic pattern of expression of a human fetal globin gene in transgenic mice. Nature 319, 685-688
Cox K H, DeLeon D V, Angerer L M and Angerer R C (1984). Detection of mRNAs in sea urchin embryos by in situ hybridisation using asymmetric RNA probes. Devl.Biol. 101, 485-502
Farace M G, Brown B A, Raschellà G, Alexander J, Gambari R, Fantoni A, Hardies S C, Hutchison C A III, and Edgell M H (1984). The mouse βh1 gene codes for the z chain of embryonic hemoglobin. J.Biol.Chem. 259, 7123-7128
Glisin V, Crkvenjakov R, Byns C (1974). Biochem. 13, 2633-2637
Hansen J N, Konkel D A and Leder P (1982). The sequence of a mouse embryonic β-globin gene. J.Biol.Chem. 257, 1048-1052
Hill A, Hardies S C, Phillips S J, Davis M G, Hutchison C A III, and Edgell M H (1984). Two mouse early embryonic β-globin gene sequences. J.Biol.Chem. 259, 3739-3747
Jahn C L, Hutchison C A III, Phillips S J, Weaver S, Haigwood N L, Voliva C F and Edgell M H (1980). DNA sequence organization of the β-globin complex in the BALB/c mouse. Cell 21, 159-168

Konkel D A, Maizel J V and Leder P (1979). The evolution and sequence comparison of two recently diverged mouse chromosomal β-globin genes. Cell 18, 865-873

Leder A, Weir L and Leder P (1985). Characterization, expression and evolution of the mouse embryonic ζ-globin gene. Mol.Cell. Biol. 5, 1025-1033

Leder P, Hansen J N, Konkel D, Leder A, Nishioka Y and Talkington C (1980) Mouse globin system: a functional and evolutionary analysis. Science 209, 1336-1342

Melton D A, Krieg P A, Rebagliati M R, Maniatis T, Zinn K and Green M R (1984). Efficient in vitro synthesis of biologically active RNA and RNA hybridization probes from plasmids containing a bacteriophage SP6 promoter. Nucl.Acids Res. 12, 7035-7056

Miura Y and Wilt F H (1969) Tissue interaction and the formation of the first erythroblasts of the chick embryo. Devl.Biol. 19, 201-211

Nishioka Y and Leder P (1979). The complete sequence of a chromosomal mouse α-globin gene reveals elements conserved throughout vertebrate evolution. Cell 18, 875-882

Peschle C, Mavilio F, Carè A, Migliaccio G, Migliaccio A R, Salvo G, Samoggia P, Petti S, Guerriero R, Marinucci M, Lazzaro D, Russo G and Mastroberardino G (1985). Haemoglobin switching in human embryos: asynchrony of ζ --> α and ε --> γ-globin switches in primitive and definitive erythropoietic lineage. Nature 313, 235-238

Popp R A, D'Surney S J and Wawrzyniak C J (1987). Changes in expression of murine α- and β-globin genes during development. In: Developmental Control of Globin Gene Expression, pp. 81-89, publ. A R Liss

Stamatoyannopoulos G, Veilh R, Al-Khatti A, Fritsch E F, Goldwasser E and Papayannopoulos Th. (1987). On the induction of fetal haemoglobin in the adult; stress erythropoieses cell cycle-specific drugs and recombinant erythropoiesis. In: Developmental Control of Globin Gene Expression, publ. A R Liss, New York (eds Stamatoyannopoulos & Nienhuis)

Whitney J B (1977). Differential control of the synthesis of two hemoglobin β chains in normal mice. Cell 12, 863-871

Wilkinson D G, Bailes J A, Champion J E and McMahon A P (1987). A molecular analysis of mouse development from 8 to 10 days post coitum detects changes only in embryonic globin expression. Development 99, 493-500

Wood W G (1976). Haemoglobin synthesis during human fetal development. Br.Med.Bull. 32, 282-284

Wood W G (1982). Erythropoiesis and haemoglobin production during development. In: Biochemical Development of the Fetus and Neonate, Ed. C T Jones, publ. Elsevier Biomedical Press, pp. 127-162

Hemoglobin Switching, Part A: Transcriptional Regulation, pages 335–342
© 1989 Alan R. Liss, Inc.

TRANSCRIPTIONAL REGULATORY FACTORS MAY CONTROL THE RELATIVE EXPRESSION OF HUMAN α1 AND α2-GLOBIN GENES IN ERYTHROLEUKEMIA CELLS

Avgi Mamalaki and Nicholas Moschonas+

Institute of Molecular Biology and Biotechnology, FORTH, +also, Dept. of Biology, U. of Crete, Heraklion 711 10, Crete, Greece.

INTRODUCTION

In human erythroid cells, the relative α2:α1 globin mRNA ratio is normally 72:28 (Liebhaber et. al. 1985). As measured in fetal liver and adult reticulocytes this ratio remains constant at least from the 10th week of gestation to adult life (Orkin and Goff, 1981). Human (mainly K562) and adult mouse erythroleukemia (MEL) cells have provided a successful model for the study of globin gene expression. Analysis of the α-globin chromatin domain has revealed that α2-gene is less active than α1 at least in the majority of K562 cells (Yagi et al. 1986). This may suggest the existence of aberrant regulatory mechanisms modifying the normal expression of the individual α1 and α2-genes in this cell line. In order to test the above hypothesis we analyzed the relative expression of α-genes both in K562 cells and MELxK562 cell hybrids (Anagnou et al., 1985) containing a single copy of human chromosome 16. With these experiments we wished to investigate the regulatory mechanisms involved in α-globin gene expression in these two cellular environments and whether this expression is under trans-control.

RESULTS

Relative activity of α1 and α2-genes in K562 cells. Assessment of α1 and α2-gene expression in K562 cells was obtained: i) by measurement of the relative abundance of α-globin transcripts in total cellular RNA and ii) by "run-on" nuclear analysis. To determine whether chemical induction can modify the relative steady state level of α1 and α2 mRNA, hemin and hydroxyurea-treated K562 cells were also tested. Induction with hemin results in a threefold increase of the rate of α-globin transcription. Hydroxyurea, although resulting in increase of globin accumulation, it does not affect the transcription rate of α-loci (Charnay and Maniatis, 1983).

Our results are consistent with the data obtained by the chromatin analysis approach regarding the α2-gene domain. In contrast to normal erythroid tissues, α2 mRNA was found substantially underrepresented (Fig. 1) and the mean α2:α1 mRNA ratio was estimated to be 8:92. In hydroxyurea-treated cells, this ratio remained unchanged. On the contrary, hemin treatment resulted in a mean ratio of 22:78, suggesting that upon transcriptional activation α2-gene is more efficient than α1.

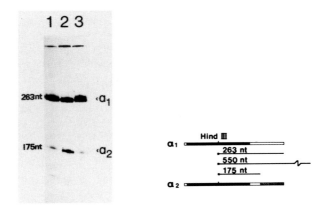

Figure 1. S1 nuclease quantitation of the relative abundance of α1 and α2-globin mRNA in K562 cells before (1) and after 50μM hemin (2) or 12μg/ml hydroxyurea (3) treatment. Total RNA was hybridized to an α1-specific cDNA probe under low stringency (Liebhaber et al. 1985). Consistent results were obtained when an α2-specific probe was used (not shown). The amount of RNA used in lane 2 is ca. one third (3 μg) of that used in lanes 1 and 3. Quantitation was done by Cerencov counting of the excised acrylamide bands. The mean α2:α1 mRNA ratios (see text) represent several repeated experiments by using either α1 or α2-specific probes.

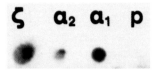

Figure 2. Dot-blot hybridization of nascent transcripts from hemin -induced K562 nuclei. ζ, a 400 bp Pst I-Hinc II fragment from pHPζ (Charnay and Maniatis, 1983); α2 and α1, DNAs from pBHα2 and pBHα1 plasmids containing the Bgl II- Hind III α-fragment from pJW101 or pRP9 (Liebhaber and Kan, 1981); p, pUC19 DNA.

    To test whether the relative expression of α-genes is controlled at the level of transcription, we examined the relative rate of α1 and α2-RNA synthesis by "run-on" nuclear transcription of hemin-treated K562 cells, where significant signals can be obtained. The α2:α1 transcription ratio was found to be 19:81 (Fig. 2), a value which is very close to that of the respective relative abundance, suggesting that the low level of α2-globin mRNA observed, is due to reduced α2-gene transcriptional activity.

    Underrepresentation of α2 mRNA to a similar extend was also determined in preliminary experiments testing HEL, OCI-M1 and OCI-M2 (Papayanopoulou et al., 1988) human erythroleukemia lines. The estimated α2:α1 mRNA ratio was ca. 10:90 for hemin or HMBA-treated HEL cells and ca. 3:97 for OCI-M1 and OCI-M2 cells induced by 5nM Ara-C and 100nM δ-Ala, respectively. "Run-on" nuclear transcription experiments with HEL cells showed comparable results with that of K562 (not shown). The relative inactivity of the α2-gene in several human erythroleukemia lines may be due either to the lack of a positive or alternative to the presence of a negative regulatory factor. The question of whether this phaenomenon shared by all lines, is due to the same molecular mechanism remains to be investigated.

    Relative expression of human α1 and α2-globin genes in MELxK562 somatic cell hybrids. To investigate whether the K562 α2:α1 mRNA ratio can be modified in *trans* after chromosomal transferring of these genes into the adult erythroid environment of MEL cells, we measured the relative abundance of α1 and α2-mRNA in three terminally differentiated MELxK562 somatic cell hybrids, i.e. 723B, 723C and 662A1. S1 nuclease protection analysis revealed that the MEL cellular environment favors expression of the α2-gene (Fig. 3); compared to the ratio obtained in K562 cells, i.e. 8:92, MEL environment resulted in a significantly shifted α2:α1 mRNA mean ratio, i.e. 52:48. Fig. 4 illustrates the effect of chemical induction and cellular environment on the relative expression of K562 α1 and α2-genes. These results provide convincing evidence that factors active in the erythroid adult environment of MEL cells can significantly modify in <u>trans</u> the relative K562 α1 and α2-gene expression.

<u>Expression of α1 and α2-globin genes after transfection in K562 cells.</u> To test whether the diverged flanking sequence immediately 3' of the α-genes participates in controlling the relative expression of the individual

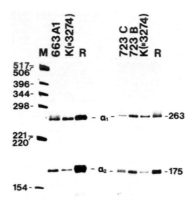

Figure 3. The relative level of α1 and α2-globin mRNA in three MELxK562 hybrids after three days of induction with 3mM HMBA and 10μM hemin. The α1-specific probe was used. Cerencov counting determined the following α2:α1 mRNA ratios: 723B, 51.5:48.5; 723C, 56.5:43.5; 663A1, 49:51. K(#3274) represents RNA from hemin-induced K562 clone #3274, the parental cell line used for the generation of the hybrids (Anagnou et al. 1985). R, reticulocyte RNA sample, exhibiting the normal α2:α1 mRNA ratio.

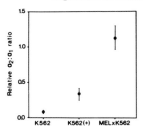

Figure 4. The relative K562 α2:α1 mRNA expression ratio upon transcriptional activation and in the adult erythroid environment of MEL cells.

α-globin genes in K562 cells, appropriate constructs containing the 1.5 Kb Pst I α1 or α2-specific fragments were transfected separately into these cells and were analyzed by S1 nuclease quantitation. To distinguish from the endogenous signal, a Cla I linker was inserted in the single Tth 111 I site of the second intron to mark the α-genes (Mα1 or Mα2 constructs). Both stable and transient expression experiments showed essentially indistinguishable Mα1 and Mα2 expression, although the stable transfection assays were limited by position effects (Table 1). Apparently, the mechanisms permitting differential expression of α-globin genes require cis-acting elements residing outside the 1.5 kb region. Hemin induction did not affect the exogenous α-globin mRNA level (data not shown).

DISCUSSION

In the present study, quantitation of the steady state α1 and α2 mRNA levels provided direct evidence that α2-gene in K562 cells, is less active than α1. As revealed by chemical induction experiments, transcriptional activation of the α-loci is sufficient for a partial relief of the aberrant α-gene expression. "Run-on" nuclear transcription suggested that the regulation of the individual α1 and α2-gene activity is largely exerted at the level of transcription. Additionally, sequences other than those included in the 1.5 kb Pst I fragment must be responsible for the differential and inducible expression of α1 and α2-genes in erythroid cells. MEL cells have been used as recipients of human chromosomes in cell fusion experiments

Table 1. Expression levels of Mα1 and Mα2-globin genes after stable or transient expression. For stable transfection, Mα1 and Mα2-genes were cloned in pLTN1 Neo expression vector (Anagnou et al. 1986). Total RNA from individual G418 resistant K562 clones and a 3' labeled 189 bp Taq I probe capable for distinguishing between exogenous (Mα) and endogenous α-transcripts, were used for S1 nuclease quantitation. Similarly, Mα1 and Mα2 cloned in pUC 19 were used for transient transfections. Transient expression signals were normalized for transfection efficiency by CAT activity after co-expression with pRSVCAT DNA.

A. Stable transfection experiment

Level of expression (% of endo. signal)	No of clones		% of clones	
	Mα1	Mα2	Mα1	Mα2
< 10%	27	19	54	56
10-35%	14	10	28	29
> 35%	9	5	18	15
TOTAL	50	34	100	100

B. Transient transfection experiment

Mα1 expression level: 34% of endogenous

Mα2 expression level: 28% of endogenous

---

to test for the presence of trans-acting factors specific for embryonic, fetal or adult globin gene expression (e.g. Anagnou et al. 1985, Papayanopoulou et al., 1985 and Baron and Maniatis, 1986). Our results demonstrate that the MEL cellular environment nearly restores the normal α2:α1 mRNA ratio, suggesting the existence of operationally positive regulatory factors in trans. Anagnou et al. (1985) have shown that MELxK562 hybrid cells induced by dimethyl formamide, produce essentially equal amounts of α-globin mRNA to that of hemin-treated K562 parental cells. We obtained consistent results bycombined treatment with HMBA and hemin. Compared to uninduced K562 cells, this suggest an ca. 3-fold increase of the level of α-globin gene expression upon MEL erythroid terminal differentiation. Thus, it can be calculated that in the induced hybrids, the augmented level of α-gene expression results from a 19 and a 1.6-fold increase of expression of α2 and α1-gene, respectively. Similarly, hemin causes in K562 cells an 8 and 2.5-fold increase of α2 and α1-gene activity, respectively. This suggest that in transcriptionally activated K562 and in terminally differentiated MEL cells, the overall increase of the α-gene expression is mainly due to activation of the α2-gene. Since ζ-gene is inactive in these hybrids (Anagnou et al. 1985) and α2 appears substantially activated, compararative chromatin structure analysis especcially of the intergenic ζ to α region, e.g. by DNase I footprinting, might reveal adjacent cis regulatory elements which could participate in the ζ to α developmental switch, in vivo. Considering the embryonic/fetal character of K562 and the adult character of MEL cells, our results demonstrate the presence of potentially stage-specific factors in erythroleukemia cells, controlling the relative expession and the level of transcription of α1 and α2-genes. It is not clear whether the same factors participate in the regulation of the α-globin gene cluster in the early stages of human development. If they do, K562 and MEL cells might offer a valuable material for their identification and isolation.

ACKNOWLEDGMENTS

This work was supported by the Hellenic Secretariat for Research and Technology. We thank Drs. N. Anagnou, T. Papayannopoulou, A. Deisseroth, T. Maniatis and M. Goossens for providing valuable cell lines, RNA samples and probes.

REFERENCES

Anagnou NP, Karlsson S, Moulton AD, Keller G, Nienhuis AW (1986). Promoter sequences required for function of the human α-globin gene in erythroid cells. EMBO J 5:121

Anagnou NP, Yuan TY, Lim E, Helder J, Wieder S, Glaister D, Marks B, Wang A, Colbert B, Deisseroth A (1985). Regulatory factors specific for adult and embryonic globin genes may govern their expression in erythroleukemia cells. Blood 65:705

Baron MH, Maniatis T (1986). Rapid reprogramming of globin gene expression in transient heterokaryons. Cell 46:591

Charnay P Maniatis T (1983). Transcriptional regulation of globin gene expression in the human erythroid cell line K562. Science 220:1281

Liebhaber SA, Cash FE, Main DM (1985). Compensatory increase in α1- globin gene expression in individuals heterozygous for the α-thalassaemia-2 deletion. J Clin Invest 76:1057

Liebhaber SA, Kan YW (1981). Differentiation of the mRNA transcripts originating from the α1 and α2-globin loci in normals and α-thalassaemics. J Clin Invest 68:439

Orkin SH, Goff SC (1981). The dublicated human α-globin genes: their relative expression as measured by RNA analysis. Cell 24:345

Papayannopoulou T, Lindsley D, Kurachi S, Lewison K, Hemenway T, Melis M, Anagnou NP, Najfeld V (1985). Adult and fetal human globin genes are expressed following chromosomal transfer into MEL cells. Proc Natl Acad Sci USA 82:780

Papayannopoulou T, Nakamoto B, Kurachi S, Tweeddale M, Messner H (1988). Surface antigen profile and globin phenotype of two new human erythroleukemia lines: characterization and interpretations. Blood 72:1029

Wilson JT, Wilson LB, de Riel JK, Villa-Komaroff L, Efstratiadis A, Forget BG, Weissman SM (1978). Insertion of synthetic copies of human globin genes into bacterial plasmids. Nucleic Acids Res 5:563

Yagi M, Gelinas R, Elder JT, Peretz M, Papayannopoulou T, Stamatoyannopoulos G, Groudine M (1986). Chromatin structure and developmental expression of the human α-globin cluster. Mol Cell Biol 6:1108

Hemoglobin Switching, Part A: Transcriptional Regulation, pages 343–357
© 1989 Alan R. Liss, Inc.

PURIFICATION OF FOUR ERYTHROID CELL PROTEINS THAT BIND THE PROMOTERS OF THE MURINE GLOBIN GENES.

Michael Sheffery, Chul G. Kim, and
Kerry M. Barnhart
DeWitt Wallace Research Laboratory and
the Graduate Program in Molecular Biology
Memorial Sloan-Kettering Cancer Center
New York, New York   10021

INTRODUCTION

The purification of ubiquitous and tissue-specific nuclear factors that interact with differentiation-specific genes is an important step towards understanding how differentiated cell lineages are established and maintained. As part of our systematic effort to identify and purify nuclear factors characteristic of cells committed to the erythroid lineage, we have purified four nuclear factors that interact strongly with the murine α-and β-globin promoters. Two factors, which we have termed α-CP1 and α-CP2, have binding sites that overlap in the α-globin CCAAT box. A third factor, termed α-IRP, interacts with two sets of sequences that form a pair of inverted repeats (IRs) located between the α-globin CCAAT and TATAA boxes (Cohen et al., 1986; Barnhart et al., 1988; Kim et al., 1988). The fourth factor, which we term EF-1 (Erythroid Factor-1), interacts with a DNA sequence element (5'-GATAAGGA-3') found near the globin genes of several species (deBoer et al., 1988; Evans et al., 1988; Super-Furga et al., 1988). Recent reports have shown that an erythroid-specific nuclear factor binds this element (deBoer et al., 1988; Evans et al., 1988; Galson and Housman, 1988; Super-Furga et al., 1988; Wall et al., 1988). Our results show that α-CP1 is comprised of seven polypeptides with M_rs between 27,000 and 38,000. These seven polypeptides are organized into a heterotypic protein complex that can be dissociated into at least two components, both of which are required for DNA binding activity. In contrast, α-CP2 is comprised of a polypeptide doublet with M_rs of 64,000 and 66,000, while a single polypeptide, with an M_r of 85,000, accounts for α-IRP binding activity. The murine erythroid-specific nuclear factor, EF-1, is comprised of a polypeptide doublet with M_rs of 18,000 and 19,000. We confirmed that the identified polypeptides correspond to each DNA binding activity either by reconstituting activity after eluting visualized proteins from sodium dodecyl sulfate (SDS)-polyacrylamide gels, or by performing UV-crosslinking experiments.

RESULTS

We have previously shown by DNase I footprinting, electrophoretic shift, column chromatography, and oligonucleotide competition experiments, that murine erythroleukemia (MEL) cell extracts contain at least three activities that bind preferentially to the α-, when compared to the β-globin promoter (Cohen et al., 1986; Barnhart et al., 1988; Kim et al., 1988). Our results identified two distinct α-globin CCAAT binding proteins (which we termed α-CP1 and α-CP2) and a third factor (α-IRP) that interacts with a pair of inverted repeat (IR) elements located between the α-globin CCAAT and TATAA boxes (Cohen et al., 1986; Barnhart et al., 1988; Kim et al., 1988). Footprinting experiments have also identified at least one erythroid-specific nuclear factor. These results have shown that nuclear extracts prepared from MEL cells, for example, contain a binding activity that protects a region near -180 on the α-globin promoter. The sequence 5'-GATAAGGA-3' is near the center of the protected region. In contrast no protection of the α-promoter is observed near -180 when footprints are performed using extracts prepared from a variety of non-erythroid cell lines or tissues (Barnhart, Kim and Sheffery, manuscript submitted). In addition, comparison of the nucleotide sequences of the mouse α- and β-globin genes suggested that the β-globin promoter might also contain a similar factor binding site, centered near -210. Indeed, this sequence is also protected by factors present in erythroid cell nuclear extracts, but not by extracts prepared from non-erythroid cells. The sequences protected upstream of the murine α- and β-globin genes are very similar to factor binding sites reported near several other erythroid-specific genes (Chambers et al., 1986; Chretien et al., 1988; deBoer et al., 1988; Evans et al., 1988; Galson et al., 1988; Kemper et al., 1987; Super-Furga et al., 1988; Wall et al., 1988). We have termed the purified murine factor that interacts with these sequences EF-1.

To purify the binding activities described above conventional chromatography was used to enrich each activity 20 to 100-fold. To enrich activities further, four different DNA sequence affinity columns were constructed based on the DNase I footprint of each enriched factor. Appropriate oligonucleotides for the affinity matricies were synthesized, hybridized, multimerized, tailed with biotinylated dUTP, and affixed to a streptavidin agarose matrix (Kadonaga and Tjian, 1986).

For affinity chromatography, fractions enriched in each binding activity were incubated with a non-specific competitor and applied to DNA sequence affinity columns in at least 100 mM KCl. After washing with at least 250 mM KCl, specific binding activity was eluted from columns with a buffer containing 1 M KCl. Column fractions were collected and assayed for specific binding activity by electrophoretic shift assays. Active fractions were pooled, salt was adjusted, non-specific competitor was added, and samples were re-applied to affinity columns. In general, 1-3 additional affinity steps were performed for each factor. After affinity chromatography, proteins were precipitated with TCA (15%), separated by electrophoresis on SDS polyacrylamide gels, and visualized by silver staining.

Fig. 1 shows that when α-CP1 was subjected to three cycles of affinity chromatography seven polypeptides with M_rs ranging from 27-38,000 were prominently enriched. To confirm that these

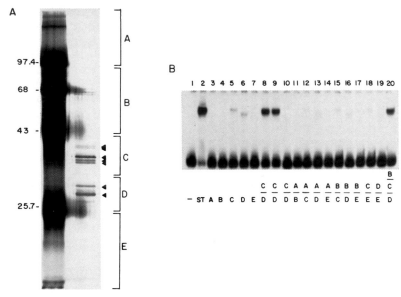

Fig. 1 Purification and confirmation of the identity of α-CP1. A. α-CP1 was purified through three cycles of affinity chromatography and run on an SDS polyacrylamide gel. B. The regions designated A-E in panel A were excised and subjected to an elution and renaturation protocol. Aliquots of renatured samples were then assayed, either alone or in the indicated combinations, for their ability to reconstitute α-CP1 binding activity in electrophoretic shift gels. Probe alone is indicated by -. ST indicates starting material.

polypeptides corresponded to α-CP1, we reconstituted the appropriate binding activity by eluting and renaturing selected polypeptides from the SDS polyacrylamide gel. As illustrated in Fig. 1, after electrophoresis of purified α-CP1, five regions of an SDS polyacrylamide gel were excised and subjected to an elution and renaturation protocol (Hatamochi et al., 1988; Kim et al., 1988). The five selected regions included polypeptides with M_rs: above 97,000 (region A); 43,000 to 97,000 (region B); 30,000 to 43,000 (region C); 25,000 to 30,000 (region D); less than 25,000 (region E). After elution and renaturation of the polypeptides contained in each region, samples were tested individually, or in various combinations, for their ability to reconstitute α-CP1 binding activity. The results showed that regions A, B and E, either alone or in combination, were devoid of any DNA binding activity. In contrast, both regions C and D contained weak amounts of DNA binding activity. Neither of the electrophoretic shifts produced by the polypeptides renatured from these two regions, however, were identical to that produced by purified α-CP1 (region C produced a slightly greater shift, whereas region D produced a slightly smaller shift). In contrast, when polypeptides eluted and renatured from regions C and D were combined, a substantial reconstitution of α-CP1 DNA binding activity was observed. Additional tests showed that only mixes containing polypeptides renatured from regions C and D (that is, only combinations including all polypeptides corresponding to affinity purified α-CP1) reconstituted the appropriate binding activity. These results confirmed our identification of α-CP1, and they suggested that α-CP1 is comprised of heterotypic protein subunits.

Fractions enriched in α-CP2 binding activity were applied to its cognate DNA sequence affinity column, and the column was developed as described above. After 3 successive passes of α-CP2 over the affinity matrix, binding activity was significantly enriched as determined by electrophoretic shift assays. Visualization of enriched proteins after separation on SDS polyacrylamide gels suggested that α-CP2 binding activity was comprised of a polypeptide doublet with M_rs of 64,000 and 66,000 (Fig. 2). We confirmed that these polypeptides account for α-CP2 binding activity by two independent means. First, after electrophoresis of purified protein, three regions of an SDS

M   1   2   3   4   5   6

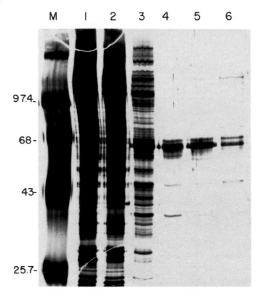

97.4-

68-

43-

25.7-

Fig. 2 Purification of α-CP2. α-CP2 was purified through 3 cycles of affinity chromatography and run on an SDS polyacrylamide gel. M: markers; lane 1: starting material; lane 2: affinity column flowthrough; lane 3: 1st affinity column; lane 4: second affinity column; lanes 5 and 6: two different 3rd affinity column samples.

polyacrylamide gel were excised and subjected to an elution and renaturation protocol, as described above. A single region that contained the polypeptides with M_rs of 64 and 66,000 was capable of reconstituting α-CP2 binding activity (Fig. 3). As an additional means to confirm that the M_r 64-66,000 polypeptides account for α-CP2 binding activity, we used UV crosslinking to transfer isotope from a radiolabelled nucleic acid probe to a specifically bound protein. The results of this experiment (Fig. 4) showed that the polypeptides with M_rs of 64-66,000 were specifically labelled by UV crosslinking to the α-globin promoter. As expected, labelling was completely abolished by addition of an unlabelled oligonucleotide containing the α-CP2 binding site. In contrast, co-incubation with an oligonucleotide containing either the α-globin IR box or a single stranded oligonucleotide containing only one strand of the α-globin CCAAT box had no effect on protein labelling by UV crosslinking. These experiments strongly suggest that the M_r 64-66,000 polypeptides account for the binding behavior of α-CP2.

α-IRP binding activity was also significantly enriched by DNA sequence affinity columns. After two passes of α-IRP over the affinity column a

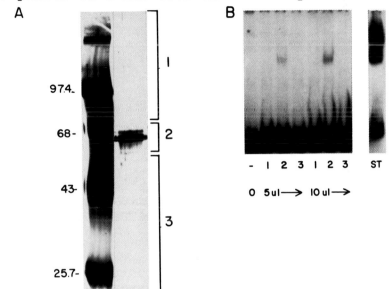

Figure 3. Confirmation of the identity of α-CP2. A. Affinity purified α-CP2 was separated on an SDS polyacrylamide gel. The three indicated regions were excised and subjected to an elution and renaturation protocol. B. Either 5 or 10 μl of sample prepared from the regions indicated in panel A was tested for α-CP2 binding activity in an electrophoretic shift assay. Starting material and probe alone lanes are indicated as in Fig. 1.

silver stained SDS polyacrylamide gel of the enriched polypeptides showed a single polypeptide

M I 2 3 4

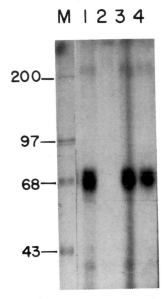

Fig. 4 UV-crosslinking also confirms identity of α-CP2. Affinity purified α-CP2 was incubated with a labelled probe containing the α-globin promoter in the absence of unlabelled competitors (lane 1) or in the presence of cold oligonucleotides specific for the α-CP2 binding site (lane 2), the α-IRP binding site (lane 3), or in the presence of single stranded DNA (lane 4). Samples were irradiated, nucleic acid was enzymatically destryed, protein was purified, separated on an SDS polyacrylamide gel, and subjected to autoradiography. Only polypeptides with M_rs of 64-66,000 were labelled in this protocol, and labelling is specifically abolished by addition of an oligonucleotide containing an α-CP2 binding site. M indicates marker proteins.

species with an M_r of 85,000 (Fig. 5). To assure that this polypeptide corresponded to α-IRP binding activity, we again performed a UV crosslinking experiment. The results obtained by UV crosslinking in the presence or absence of specific oligonucleotide competitors confirmed that the

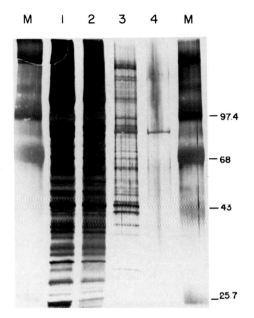

Fig. 5. Purification of α-IRP. α-IRP was subjected to DNA sequence affinity chromatography, and enriched polypeptides were separated on an SDS polyacrylamide gel. M: marker proteins; lane 1: starting material; lane 2: 1st affinity flowthrough; lane 3: 1st affinity column 1M salt cut; lane 4: 2nd affinity column 1M salt cut.

protein with an M_r of 85,000 coincides with α-IRP binding activity (Fig. 6).

EF-1 binding activity was also purified by DNA affinity chromatography. After two affinity steps, two sets of polypeptide doublets having M_rs of 36 and 37,000 and 18, and 19,000 were prominently enriched. Two additional cycles of affinity chromatography under more stringent conditions, however, yielded only the polypeptide doublet with M_rs of 18 and 19,000 (Barnhart, K.M, C.G. Kim, and M. Sheffery, manuscript submitted). To confirm that the polypeptides with M_rs of 18 and 19,000 account

for EF-1 binding activity we reconstituted the
appropriate activity by eluting and renaturing

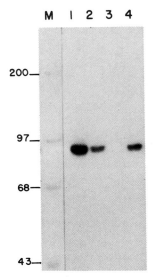

Fig. 6. UV-crosslinking confirms identification of
of α-IRP. Affinity purified α-IRP was subjected to
a UV-crosslinking protocol, as described in Fig. 4.
M: markers; lane 1: α-IRP incubated in the absence
of competitor; lane 2: incubation with an
oligonucleotide containing a strong α-CP2 binding
site; lane 3: incubation with an oligonucleotide
containing an α-IRP binding site; lane 4:
incubation with single stranded DNA.

polypeptides from SDS polyacrylamide gels. After
electrophoresis of purified EF-1 binding activity,
five regions of an SDS polyacrylamide gel were
excised and subjected to an elution and
renaturation protocol. Renatured samples were
tested for their ability to reconstitute EF-1
binding activity. A single region of the gel, which
contained the polypeptides with M_rs of 18,000 and
19,000, reconstituted EF-1 binding activity,
strongly suggesting that these polypeptides
correspond to EF-1 (Barnhart, K.M, C.G. Kim, and M.
Sheffery, manuscript submitted).

DISCUSSION

In an attempt to gain further insight into the nuclear factors that interact with globin genes, we have used DNA sequence affinity chromatography to purify four factors that interact with the murine α- and β-globin promoters. Three factors (α-CP1, α-CP2, and α-IRP) interact selectively with the murine α-globin promoter. A fourth factor, EF-1, is erythroid-cell specific, and interacts with a sequence (5'-GATAAGGA-3') found near many erythroid cell specific, or erythroid cell induced promoters (Chambers et al., 1986; Chretien et al., 1988; deBoer et al., 1988; Evans et al., 1988; Galson et al., 1988; Kemper et al., 1987; Super-Furga et al., 1988; Wall et al., 1988).

Our results show that α-CP1 is comprised of seven polypeptides with M_rs that range from 27 to 38,000. The identity of α-CP1 was confirmed by experiments designed to reconstitute α-CP1 binding activity by eluting and renaturing specific polypeptides from SDS polyacrylamide gels. These results confirmed that polypeptides with M_rs from 27 to 38,000 are required for α-CP1 binding activity. In addition they showed that α-CP1 binding activity is fully reconstituted only when at least two distinct sets of polypeptides are combined. Together, these results suggest that the α-globin CCAAT box interacts strongly with a factor, which we have termed α-CP1, comprised of a heterotypic protein complex. Our observations are consistent with several recent reports suggesting that heterologous protein complexes might be typical of eucaryotic CCAAT binding proteins (Chodosh et al., 1988; Hatamochi et al., 1988). We also note that we have previously shown (Barnhart et al., 1988) that α-CP1 interacts only weakly with the adenovirus origin of replication, a strong binding site for another identified CCAAT factor termed CTF/NF-1 (Jones et al., 1987). The inability of α-CP1 to bind the adenovirus origin of replication, together with our purification data, strongly suggests that α-CP1 is not identical to CTF/NF-1.

The second purified factor, α-CP2, binds largely upstream of, but overlaps with, the α-globin CCAAT box (Barnhart et al., 1988). Indeed, we have previously shown that α-CP1 and α-CP2 have overlapping binding sites, and that these two factors can compete for occupancy of the α-globin CCAAT box in vitro (Barnhart et al., 1988). Affinity purification of α-CP2 revealed a polypeptide doublet with M_rs of 64 and 66,000. Both UV crosslinking and renaturation of activity from SDS polyacrylamide gels confirmed that these polypeptides are responsible for α-CP2 binding activity.

A third purified factor interacts with two sequences that form a pair of inverted repeats (IRs) between the CCAAT and TATAA boxes (Barnhart et al., 1988). We termed this factor α-IRP. α-IRP is comprised of a single polypeptide with an M_r of 85,000. UV crosslinking experiments confirmed that this polypeptide accounts for α-IRP binding activity. We note here that because the α-globin IR boxes contain GC rich sequences reminiscent of SP1 binding sites, we have previously tested the ability of α-IRP to bind the GC-boxes in the 21 bp repeats of the SV40 early promoter (Barnhart et al., 1988). We found that α-IRP binds as well to the α-promoter as it does to the SV40 early promoter, suggesting that this erythroid factor might be related to the HeLa cell derived transcription factor SP1 (Briggs et al., 1986). In addition to many chromatographic differences, however, the observed mobility of purified α-IRP in SDS polyacrylamide gels is distinct from that of purified SP1 (Briggs et al., 1986). Preliminary experiments also suggest that α-IRP footprints the SV40 early promoter in a manner distinctive from SP1. While there may be a family of SP1-like polypeptides, the simplest interpretation of our data is that α-IRP has features that distinguish it from HeLa cell derived SP1.

The fourth purified activity, EF-1, is detected only in nuclear extracts prepared from erythroid cells. EF-1 binding activity is accounted for by a polypeptide doublet with M_rs of 18 and 19,000 (Barnhart, Kim and Sheffery, manuscript submitted). The identity of EF-1 was confirmed by reconstituting the appropriate binding activity after eluting and renaturing purified polypeptides from SDS polyacrylamide gels (Barnhart, K.M, C.G. Kim, and M. Sheffery, manuscript submitted).

The absence of EF-1 binding activity in cells of non-erythroid lineages suggests that this factor might play a crucial role in erythroid development and differentiation, a notion supported by several recent findings. For example, the DNA sequence element to which EF-1 binds is found near the globin genes of several species (deBoer et al., 1988; Evans et al., 1988; Galson and Housman, 1988; Kemper et al., 1987; Super-Furga et al., 1988; Wall et al., 1988). Moreover, the same sequence is found near the promoters of other genes whose expression is induced in erythroid cells, such as the glutathione peroxidase promoter (Chambers et al., 1986) and the erythrocyte specific promoter of the porphobilinogen deaminase gene (Chretien et al., 1988). In addition, mutation of the region in the chicken β-globin 3' enhancer that binds an obviously related erythroid-specific factor results in a profound loss of enhancer activity, as measured in transient expression studies (Reitman et al., 1988). Together, these results suggest a potentially important role for EF-1 in mediating erythroid cell differentiation.

While we have purified four factors, including one erythroid-specific factor, that interact strongly with the α- and β-globin promoters (and other nearby sequences) we emphasize that the function these proteins play in regulating globin gene expression during erythroid cell differentiation is poorly understood. Although all of the factors we have purified affect α-globin gene transcription in vitro when added to either HeLa or MEL cell nuclear extracts (M. Sheffery, unpublished results), how these factors cooperate to regulate α-globin gene transcription in vivo is unknown. Our ability to purify each factor should expedite the cloning and production of factor-specific antisera that will be required to obtain deeper insights into the function each factor plays in regulating erythroid cell differentiation.

ACKNOWLEDGEMENTS

This work was supported, in part, by Public Health Service grants CA-31768 and CA-08748 from the National Cancer Institute, grant DK-37513 from the NIH, and by the Businessmen's Assurance Fund. The technical assistance of Kendra Dean is gratefully acknowledged.

REFERENCES

Barnhart KM, Kim CG, Banerji SS, Sheffery M (1988). Identification and characterization of multiple erythroid cell proteins that interact with the promoter of the murine α-globin gene. Mol Cell Biol **8**: 3215-3226.

Briggs MR, Kadonaga JT, Bell SP, Tjian R (1986). Purification and biochemical characterization of the promoter-specific transcription factor, SP1. Science **234**: 47-52.

Chambers I, Frampton J, Goldfarb P, Affara N, McBain W, Harrison PR (1986). The structure of the mouse glutathione peroxidase gene: the selenocycteine in the active site is encoded by the 'termination' codon TGA. EMBO J **5**:1221-1227.

Chodosh LA, Baldwin AS, Carthew RW, Sharp PA (1988). Human CCAAT-binding proteins have heterologous subunits. Cell **53**: 11-24.

Chretien S, Dubart A, Beaupain D, Raich N, Grandchamp B, Rosa J, Goossens M, Romeo P.-H (1988). Alternative transcription and splicing of the human porphobilinogen deaminase gene result either in tissue-specific or in housekeeping expression. Proc Natl Acad Sci USA **85**: 6-10.

Cohen RB, Kim CG, Sheffery M (1986). Partial purification of a nuclear protein that binds to the CCAAT box of the mouse α_1-globin gene. Mol Cell Biol **6**: 821-832.

deBoer E, Antoniou M, Mignotte V, Wall L, Grosveld F (1988). The human β-globin promoter; nuclear protein factors and erythroid-specific induction of transcription. EMBO J **7:** 4203-4212.

Evans T, Reitman M, Felsenfeld, G (1988). An erythrocyte-specific DNA binding factor recognizes a regulatory sequence common to all chicken globin genes. Proc Natl Acad Sci USA **85:** 5976-5980.

Galson DL, Housman DE (1988). Detection of two tissue-specific DNA binding proteins with affinity for sites in the mouse β-globin intervening sequence 2. Mol Cell Biol **8:** 381-392.

Hatamochi A, Golumbek PT, van Schaftingen E, de Crombrugghe B (1988). A CCAAT DNA binding factor consisting of two different components that are both required for DNA binding. J Biol Chem **263:** 5940-5947.

Jones KA, Kadonaga JT, Rosenfeld PJ, Kelly TJ, Tjian R (1987). A cellular DNA-binding protein that activates eukaryotic transcription and DNA replication. Cell **48:** 79-89.

Kadonaga JT, Tjian R (1986). Affinity purification of sequence-specific DNA binding proteins. Proc Natl Acad Sci USA **83:** 5889-5893.

Kemper B, Jackson PD, Felsenfeld G (1987). Protein binding sites within the 5' DNase I-hypersensitive region of the chicken α^D-globin gene. Mol Cell Biol **7:** 2059-2069.

Kim CG, Barnhart KM, Sheffery M (1988). Purification of multiple erythroid cell proteins that bind the promoter of the α-globin gene. Mol Cell Biol **8:** 4270-4281.

Reitman M, Felsenfeld G (1988). Mutational analysis of the chicken β-globin enhancer reveals two positive-acting domains. Proc Natl Acad Sci USA **85:** 6267-6271.

Super-Furga G, Barberis A, Schaffner G, Busslinger M (1988). The -117 mutation in Greek HPFH affects the binding of three nuclear factors to the CCAAT region of the gamma-globin gene. EMBO J **7:** 3099-3107.

Wall L, deBoer E, Grosveld F (1988). The human β-globin gene 3' enhancer contains multiple binding sites for an erythroid-specific protein. Genes Dev **2:** 1089-1100.

Hemoglobin Switching, Part A: Transcriptional Regulation, pages 359–366
© 1989 Alan R. Liss, Inc.

REGULATED EXPRESSION OF THE ERYTHROID-SPECIFIC-PROMOTER OF THE HUMAN PBG-D GENE DURING ERYTHROID DIFFERENTIATION

Paul-Henri Romeo, Vincent Mignotte, Natacha Raich, Anne Dubart, Denise Beaupain, Marc Romana, Claude Chabret, Lee Wall[*], Ernie deBoer[*], Frank Grosveld[*]

INSERM U 91, Hôpital H. Mondor, 94010 Créteil France
[*] Laboratory of Gene Structure and Expression, National Institute for Medical Research, The Ridgeway, Mill Hill, London NW7 1AA U.K.

INTRODUCTION

The final cellular product of the erythroid differentiation, the erythrocyte, contains massive amounts of hemoglobin. Hemoglobin biosynthesis needs the coordinated production of globin chains and heme molecules. The synthesis of the globin chains primarily results from the transcriptional activation of the globin genes (Charnay et al., 1984; Wright et al., 1984) and the enhanced heme production is ensured by the activation of the enzymes involved in its biosynthesis. For at least three of those enzymes, this activation is partly accounted for by an increased transcription of their genes (Raich et al., 1986; Romeo et al., 1986; Schoenhaut and Curtis, 1986). Whether globin genes and genes coding for the enzymes of the heme biosynthetic pathway are transcriptionally regulated by similar mechanisms is still unknown.

We have focused our work on the expression of the human gene coding for Porphobilinogen deaminase (PBG-D; E.C.4.3.1.8), the third enzyme of the heme biosynthetic pathway. This gene is 10 kb long and split into 15 exons (Chretien et al., 1988). It contains two overlapping transcription units each with its own promoter. The upstream promoter is active in all cell types while the downstream promoter is erythroid specific. After differential splicing, an ubiquitous or an erythroid mRNA, which differ exclusively at their 5' ends (Grandchamp et al.,

1987; Raich et al., 1986) are produced **(Fig. 1)**.

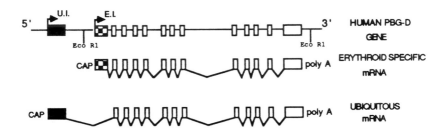

Fig. 1: Structure of the human PBG-D gene and of its two transcripts. U.I. and E.I. indicate the ubiquitous and erythroid initiation sites, respectively.

Comparison between the human β-globin gene promoter and the erythroid promoter of the PBG-D gene has suggested a modular organization [5'...CAAC box...6bp... CAAT box... TATA box...CAP...3'] which might account for coregulation of those two genes during erythroid differentiation. We have studied the erythroid-specific promoter of the human PBG-D gene by transfection into MEL cells and by DNAase I footprinting and gel retardation. We showed that this promoter was inducible during MEL cell differentiation and defined a 321 bp DNA region which was sufficient for correct regulation. This DNA fragment contained three binding sites for the erythroid-specific factor NF-E1 and one site for a second erythroid-specific factor which we named NF-E2. Site-directed mutagenesis indicated that cooperativity between NF-E1 and a yet uncharacterized factor accounted for correct regulation of the PBG-D erythroid promoter during MEL cell differentiation.

Deletion analysis of the erythroid promoter of the PBG-D gene

We have previously shown (Raich et al., submitted) that a hybrid gene containing the PBG-D erythroid promoter (-714 to +78) fused to the Herpes Simplex Virus thymidine kinase (HSV tk) coding sequence was correctly expressed and regulated when introduced into murine erythroleukemia (MEL) cells. This result suggested that structural features within the PBG-D erythroid promoter were responsible for its activation during erythroid differentiation. To delimit the sequences involved in the regulation of the PBG-D erythroid promoter, we performed sequential deletions of this promoter. As indicated in **Figure 2**, the - 243 mutant was correctly regulated during MEL cell differentiation whereas the - 112 mutant was not. These results indicate that an element, located between - 243 and - 112 relative to the erythroid start site, was involved in the erythroid regulation of this promoter.

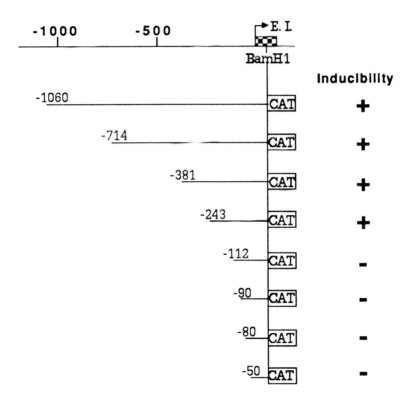

Fig. 2: Effect of 5' deletions on the regulation of the PBG-D erythroid promoter during MEL cell differentiation. E.I. indicates the PBG-D erythroid initiation site.

Analysis of the DNA binding factors which recognized the PBG-D erythroid promoter

A promoter fragment spanning nucleotides - 243 to + 78 relative to the erythroid transcription start site was analyzed for the binding of nuclear factors in vitro. Analysis of the sequences within the footprinted regions revealed three putative erythroid-specific binding sites [- 180; - 70 and + 45 regions] to contain the consensus motif C/A Py T/A ATC T/A Py which had been shown to bind the erythroid-specific protein NF-E1 (Wall et al., 1988; de Boer et al., 1988; Evans et al., 1988). In addition, a motif around - 160 was able to bind different proteins in non erythroid and erythroid cells. An erythroid-specific factor, which we called NF-E2, competed for binding to this - 160 sequence with the AP1 complex (Angel et al., 1987). Finally, ubiquitous motifs (see **Fig. 3**) were present in the PBG-D erythroid promoter. Some, like CAC binding proteins, also bind to the β-globin promoter while others, like NF-U1 had not been identified either by competition or by comparison with known binding sites.

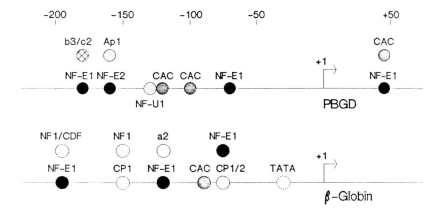

Fig. 3: Comparison of protein-DNA interactions on the human β globin and PBG-D erythroid promoters. Numbers indicates promoter coordinates relative to the transcription initiation site (For details see Mignotte et al., 1988).

The – 70 NF-E1 binding site is necessary but not sufficient for correct regulation of the PBG–D erythroid promoter

To assess the function of those three NF-E1 binding sites found in the PBG-D erythroid promoter, we performed a site directed mutagenesis on each of those sites. The recombinant plasmids were transfected into MEL cells. As shown in **Figure 4,** the suppression of the – 180 or + 45 NF-E1 binding sites did not alter the regulation of the PBG-D erythroid promoter whereas the suppression of the – 70 NF-E1 binding site completely abolished the regulation of the PBG-D promoter during MEL cell differentiation.

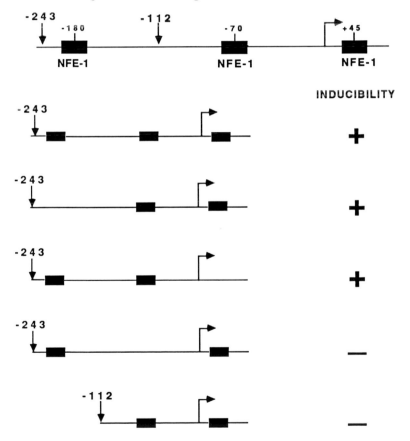

Fig. 4: The – 70 NE-E1 site is necessary but not sufficient for correct regulation of the PBG-D erythroid promoter.

As we have shown that the - 112 mutant, which contained the - 70 NF-E1 binding site and the CAC box, was not regulated during MEL cell differentiation, those results suggested that cooperativity between NF-E1 and a yet uncharacterized factor was necessary for correct regulation of the PBG-D erythroid promoter during MEL cell differentiation.

CONCLUSION

Transcription of the globin genes and genes coding for enzymes of the heme biosynthetic pathway increases during erythroid differentiation. It is therefore probable that a common mechanism is responsible for this coordinated expression. Our previous studies have shown that the human gene encoding porphobilinogen deaminase, the third enzyme of the heme biosynthetic pathway, has two promoters. The upstream one is active in all cell types whereas the downstream promoter is active only in erythroid cells. The analysis of the PBG-D erythroid promoter might therefore give insights into the mechanisms responsible for the inducible erythroid expression and coregulation of different sets of genes.

The results presented here have shown that the PBG-D erythroid promoter and the human β globin promoter might be coregulated at the transcriptional level. We have shown that NF-E1, an erythroid specific transacting factor, had three binding sites in the PBG-D erythroid promoter. This factor has been shown to bind multiple sites in the human β globin promoter and enhancer (Wall et al., 1988; de Boer et al., 1988) as well as in the chicken β globin enhancer (Evans et al., 1988). Thus, NF-E1 might account for coregulation of erythroid specific genes during erythroid differentiation.

We have also shown, using site directed mutagenesis, that another factor(s) which binds to the promoter between - 243 and - 112 was necessary for correct regulation of the PBG-D erythroid promoter during MEL cell differentiation. An attractive candidate is another erythroid-specific factor, which we called NF-E2. This factor recognized a sequence around - 160 which contains a consensus motif for the non-erythroid-specific complex AP1 (Angel et al., 1987). Interestingly, this same sequence is found in a functionally important region of the chicken β globin enhancer (Reitman and Felsenfeld, 1988). Thus NF-E2 might also be involved in the coregulation of erythroid specific genes.

REFERENCES

Angel P, Imagawa M, Chiu R, Stein B, Imbra RS, Rahmsdorf HS, Uonat C, Herrlich P and Karin M (1987) Phorbol ester-inducible genes contain a common cis element recognized by a TPA-modulated trans-acting factor. Cell 49:729-739.

Charnay P, Treisman R, Mellon P, Chao M, Axel R and Maniatis T (1984) Differences in human α- and β-globin gene expression in mouse erythroleukemia cells: the role of intragenic sequences. Cell 38:251-263.

Chretien S, Dubart A, Beaupain D, Raich N, Grandchamp B, Rosa J, Goossens M and Romeo PH (1988) Alternative transcription and splicing of the human porphobilinogen deaminase gene result either in tissue-specific or in housekeeping expression. Proc Natl Acad Sci USA, 85:6-10.

deBoer E, Antoniou M, Mignotte V, Wall L and Grosveld F (1988) The human β-globin gene promoter: nuclear protein factors and erythroid-specific induction of transcription. Embo J in press.

Evans T, Reitman M and Felsenfeld G (1988) An erythrocyte-specific DNA-binding factor recognizes a regulatory sequence common to all chicken globin genes. Proc Natl Acad Sci USA 85:5976-5980.

Grandchamp B, de Verneuil H, Beaumont C, Chretien S, Walter O and Nordmann Y (1987) Tissue specific expression of porphobilinogen deaminase. Two isoenzymes from a single gene. Europ J Biochem 162:105-110.

Mignotte V, Wall L, deBoer E, Grosveld F and Romeo PH (1988) Two tissue-specific factors bind the erythroid promoter of the human porphobilinogen deaminase gene. Nucl Acids Res (in press).

Raich N, Romeo PH, Dubart A, Beaupain D, Cohen-Solal M and
  Goossens M (1986) Molecular cloning and complete primary
  sequence of human erythrocyte porphobilinogen deaminase.
  Nucl Acids Res 14:5955-5968.

Reitman M and Felsenfeld G (1988) Mutational analysis of
  the chicken β-globin enhancer reveals two positive-acting
  domains. Proc Natl Sci USA 85:6267-6271.

Romeo PH, Raich N, Dubart A, Beaupain D, Pryor M, Kushner
  J, Cohen-Solal M and Goossens M (1986) Molecular cloning
  and nucleotide sequence of a complete human uroporphyri-
  nogen decarboxylase cDNA. J Biol Chem 261:9825-9831.

Schoenhaut DS and Curtis PJ (1986) Nucleotide sequence of
  mouse 5-aminolevulinic acid synthase cDNA and expression
  of its gene in hepatic and erythroid tissues. Gene
  48:55-63.

Wall L, deBoer E and Grosveld F (1988) The human β-globin
  gene 3' enhancer contains multiple binding sites for an
  erythroid-specific protein. Genes and Devel 2:1089-1100.

Wright S, Rosenthal A, Flavell RA and Grosveld FG (1984)
  DNA sequences required for regulated expression of
  β-globin genes in murine erythroleukemia cells. Cell
  38:265-273.

Index